ASTRONOMY and CEREMONY in the PREHISTORIC SOUTHWEST

edited by

John B. Carlson and W. James Judge

Papers of the Maxwell Museum of Anthropology

Number 2

Library of Congress Cataloging-in-Publication Data

Astronomy and ceremony in the prehistoric Southwest.

(Papers of the Maxwell Museum of Anthropology ; no. 2)
Majority of papers presented at a symposium Astronomy and ceremony in the prehistoric Southwest, sponsored by the Maxwell Museum of Anthropology at the Univeristy of New Mexico, Oct. 12-14, 1983.
Bibliography: p.
1. Indians of North America--Southwest, New--Astronomy --Congresses. 2. Indians of North America--Southwest, New--Rites and ceremonies--Congresses. 3. Astronomy, Prehistoric--Southwest, New--Congresses. 4. Indians of North America--Southwest, New--Antiquities--Congresses. 5. Southwest, New--Antiquities--Congresses. I. Carlson,

John B., 1945- . II. Judge, W. James (William James) III. Series.
E78.S7A77 1987 979'.01 87-5592
ISBN 0-912535-03-2

CONTENTS

Contents (continued)

Contents (continued)

ILLUSTRATIONS

Illustrations (continued)

TABLES

Foreword

Marsha C. Bol

The majority of the articles in this collection were presented as papers at a symposium entitled "Astronomy and Ceremony in the Prehistoric Southwest," sponsored by the Maxwell Museum of Anthropology at the University of New Mexico and held on October 12-14, 1983. The symposium developed in conjunction with a major exhibition entitled "The Chaco Phenomenon," which was organized by the Maxwell Museum and which later traveled to major museums throughout the country. "The Chaco Phenomenon" focused on the prehistoric Anasazi occupation of Chaco Canyon in northwestern New Mexico, with its complex and widespread social and economic system.

The occasion of the exhibition offered an opportunity to explore more fully one of the issues related to the social and ritual systems of prehistoric Chaco Canyon and of the Anasazi. In view of the Maxwell Museum's ongoing commitment to public education in areas relevant to the Southwest, and in particular to encourage public awareness of and responsibility toward anthropological resources, a two and one-half day symposium exploring the topic of Southwestern archaeoastronomy, or the investigation of ancient astronomical systems, seemed warranted. W. James Judge, then director of the Chaco Center, a research division of the National Park Service, was aware of the need for public education regarding Southwest astronomical sites and supported the project; he was involved in planning the symposium from its inception and served as its chairman.

Archaeoastronomers have investigated the Southwest intensively in the last decade, focusing frequently on Chaco Canyon. With the release of the film "The Sun Dagger" to public television in 1982, both Southwest archaeoastronomy and the Sun Dagger site on Fajada Butte in Chaco Canyon were brought to the attention of a wide audience. The public responded with keen interest --so avidly, in fact, that the Fajada Butte site has since been closed to visitation both for reasons of public safety and to preserve the site itself.

The purpose of the symposium was to present and synthesize recent evidence concerning archaeoastronomy in the Southwest and to discuss the role of astronomical observations in the ceremonial life of prehistoric people. Research at Chaco Canyon provided the perspective and the opportunity both to examine the evidence in a regional context and to compare the results with those from other prehistoric Southwestern culture areas.

The symposium was the first public forum to be held on the topic of Southwest archaeoastronomy. The approach was innovative, as is the discipline of archaeoastronomy itself. A group of scholars from many backgrounds, including archaeologists, artists, astronomers, cultural anthropologists, historians, and physical and political scientists, were asked to present new research and, at the same time, to debate some of the basic issues and differing viewpoints in open discussion. This interchange of ideas is critical to the process of investigating astronomy as an important component of prehistoric ceremony because this is by no means a field in which the experts concur on the various issues. A summary of the issues and central themes presented both in the papers delivered on the first day and in the series of panel discussions directed by John Carlson on the second day of the symposium was prepared by M. Jane Young and appears in Chapters 10 and 13.

Special thanks are due to the staff at the Maxwell Museum. J. J. Brody, then museum director and curator of "The Chaco Phenomenon," was both supportive and trusting in this endeavor. Garth Bawden, current director of the museum, has been very helpful in bringing the results of the symposium to publication. The book was typeset and copy edited by Lynne Sebastian and June-el Piper and pasted up by Jerry Livingston. Mr. Livingston also prepared the original artwork that appears on the cover. The symposium was made possible by a grant from the Maxwell Museum Association. Special thanks are also given to Blair Darnell, program chairperson, and to association presidents Maxine Friedman and Marshall Nason. The many volunteers from the association membership, organized by Peggy Cavett-Walden, made it possible for a small museum to host a large symposium, and credit is due to all of these people for the success of this effort.

Chapter 1

Archaeology and Astronomy:

A View from the Southwest*

W. James Judge

Recently I asked a friend of mine, an archaeologist whose opinion I respect very much, what he thought of the subject of archaeoastronomy. He said, "Well, it's one-half archaeology and one-half astronomy, but unfortunately, much of it has turned out to be sort of halfastroarchaeology." This, I am afraid, is a perception of the field shared by a number of archaeologists, a perception that conditions the character of this essay.

I would like to discuss the relationship between archaeology and astronomy, as they are integrated in the emerging field of archaeoastronomy. You must realize that my talk is presented from the perspective of an archaeologist who knows something about the subject of archaeology--at least Southwestern archaeology--but virtually nothing about the subject of astronomy. (I am one of the few people who has trouble finding the Milky Way, let alone the Big Dipper.) Thus, I ask you to please understand and forgive my biases.

This chapter will offer some thoughts on the distinction between archaeology and archaeoastronomy (as well as between archaeologists and archaeoastronomers), followed by a review of some of the work we have done in Chaco Canyon. I will conclude by offering suggestions about ways in which research in Chaco can help to integrate the two disciplines, illustrating the challenge thereby posed to those who work in the field of archaeoastronomy.

I would like to preface all of this by stating some of the basic assumptions that will underlie this discussion. First, I feel that the field of archaeoastronomy has more ultimate relevance to archaeology than to astronomy, and thus should make its home in the camp of the former. Though prehistoric astronomy may be of interest relative to the history of astronomy, I doubt that the key to the origin of the universe will ultimately be found in Pueblo Bonito. Archaeology, on the other hand, can benefit considerably from an understanding of the role of astronomy in prehistoric cultures.

Second, I assume that like it or not, a dichotomy of considerable magnitude exists between archaeology and archaeoastronomy at this time, in this country. Otherwise, why would my archaeological colleague

* This chapter is a revised version of the keynote address given by Dr. Judge at the symposium.

have made the remark he did? My third assumption is that this gap can be, and most certainly should be, bridged--if we work at it. I am convinced that symposia such as the one reported here, which offer opportunities for archaeologists, astronomers, cultural anthropologists, and others with similar interests to get together and talk to each other, are of great value in providing the emerging field of archaeoastronomy with a more fundamentally anthropological foundation from which to develop.

Although I do not know for sure, I would bet that most of those who now consider themselves archaeoastronomers were actually trained in astronomy, rather than in archaeology or, more correctly, in anthropology. In the United States one cannot get a degree in archaeology per se. Archaeology is subsumed within the larger field of anthropology, which is the study of human behavior. This intellectual background, of course, conditions how archaeologists go about their business. My guess is that archaeoastronomers whose degrees are in anthropology are the exception rather than the rule with regard to formal training--Jonathan Reyman, a participant in this symposium, is one of the exceptions.

Dominance of the field of archaeoastronomy by astronomers obviously conditions both the character of the research undertaken and the interpretations derived and does not necessarily foster integration of the discipline with anthropological archaeology. Please understand that I am neither blaming nor belittling the astronomers--in fact, were it not for their interest, the field of archaeoastronomy would have shown very little progress since its inception. It is, however, this distinct intellectual background of archaeoastronomy that I wish to address. Perhaps the best way to begin is to compare astronomy with archaeology.

First, let's consider the public perception of astronomy. When one thinks of astronomy, what comes to mind? Galaxies, black holes, quarks, Carl Sagan, warp factor one, the starship Enterprise . . . things like that. Now, think of archaeology and what comes to mind? Dirt, trash, refuse, garbage, potsherds, broken rocks, and generally, dinosaur bones (which aren't even in the domain of the archaeologist). Now, think of archaeoastronomy and what comes to mind? Sun daggers, supernova, calendars, mystery, Carl Sagan, glamour, sun<u>rise</u> . . . the thrill of discovery. Now think of archaeology again and what comes to mind? More garbage, more pottery, more rocks, dirt, dust, and sun<u>burn</u> . . . the agony of defeat.

I'm not complaining, you understand. We archaeologists know our place, are content with our lot, and generally enjoy what we do. It's just that we get no respect. I have discussed this elsewhere and have called it the "Rodney Dangerfield" syndrome in archaeology. I am, of course, being facetious in emphasizing our paranoia, but there is a point here, and it has to do with the public perception of what we as archaeologists do for a living and how that relates to the things that are more likely to capture the attention and the interest of the public--things like medicine wheels, stone monoliths, corner windows, kiva niches, and strange petroglyphs --things, incidentally, that are the lifeblood of the archaeoastronomers.

Archaeology deals with, actually archaeology <u>must</u> deal with, the mundane, the commonplace, the discarded remnants of human behavior. Therein lies the difference between it and astronomy, and perhaps the key to achieving the integration of the two disciplines. Let me try to explain.

As I noted earlier, archaeologists are trained in anthropology, and thus they are primarily interested in human behavior--specifically, in the processes of change through time in human behavior and, even more specifically, in why such changes take place. I mentioned before that archaeologists deal with trash--we enjoy it and like to work with it. You may be familiar with a project in Tucson, Arizona, in which archaeologists are studying modern human trash--actually going through people's garbage in that fair city to see what they throw away. Dr. Rathje, the project director, has noted that while all archaeologists study garbage, his data were just a little fresher than most. What does this have to do with archaeoastronomy? Please bear with me.

The purpose of the garbage project is to compare what the researchers call "front door" and "back door" behavior in Tucson--in other words, to compare what people say they do with what they actually do. What is the relevance of this to archaeology? Basically, Rathje is trying to get at what we call the "material correlates" of modern human behavior. He wants to understand the relationship between how humans actually behave and what they actually leave in the archaeological record. Understanding this will help archaeologists to comprehend the relationship between what we find as we dig prehistoric trash mounds and other discarded remains and how the people actually behaved in the past. Incidentally, as an interesting aside, Rathje has found that in Tucson people discard between 7 and 14 percent, by weight, of their purchased food. This would come to about $11 billion annually for the nation as a whole, or enough to feed the entire population of Canada for a year.

But I digress. Back to the archaeological correlates of human behavior. This quest to determine what conditions the formation of the archaeological record now constitutes a primary thrust of archaeological research in this country. Theoretical and empirical attempts to verify the inferred function of archaeological manifestations, sometimes called middle-range theory, are among the foremost issues confronting archaeologists today. For example, take something as simple as an excavated room in a prehistoric pueblo. How are we to interpret the function of that room? If we call it a habitation room, how are we to know that it indeed functioned in that context? What are the material correlates of domestic living, and how would they differ from the material remains resulting from other possible functions of the room? Is a room without a hearth really a storage room? Or how about a room with heating pits rather than hearths? Was it a living room, a storage room, or both? Sometimes it seems as though we have not progressed very far in interpreting the archaeological record if we cannot even discern something as basic as room function.

I am sure you can appreciate the importance of understanding things as fundamental as this. How, for example, can we reconstruct past population size without being able to determine which rooms served primarily for habitation, which for storage, and for what portion of the year they so served? Thus we see the reasons behind the current emphasis on understanding the material correlates of human behavior.

Let me try to give you an example from the field of archaeoastronomy. Take, for instance, the so-called supernova pictograph located in Chaco Canyon near Peñasco Blanco, a site that is two quite dusty but otherwise very beautiful miles west of Pueblo Bonito. Some astronomers have suggested that this pictograph is an Anasazi record of the Crab Nebula supernova, which was recorded by Chinese astronomers in AD 1054. The time is right, since the Chaco system peaked between AD 1050 and 1130. Some archaeologists, on the other hand, have suggested that this pictograph is nothing more than a prehistoric sunwatcher's station, a phenomenon not uncommon among Pueblo peoples and nothing to get very excited about.

Whom should we believe? Well, for one thing, I agree fully with Zeilik's approach (Chapter 3) to the question, which is to try to test the sunwatcher hypothesis. Rather than arguing about it, we should go out and actually observe the sun from the location of the pictograph to see whether the resultant observations make any sense whatsoever, given what we know about solar observations among the modern Pueblos. To my knowledge, Zeilik is the first one to suggest this. I don't know whether he has done it yet, or if so, what he found out. But it is certainly worth a try to determine, in effect, whether this site exhibits the material correlates of sunwatching behavior. That is the anthropological perspective.

Let me take this opportunity to point out something else about the supernova pictograph, something with which many of you might not agree. If it were determined that this site is a sunwatcher's station, that would be interesting, but it would not change our interpretation of the Chaco Phenomenon very much. If, on the other hand, it were determined that this rock art panel is a record of the supernova of 1054, a record of a unique celestial event, what would that mean? Little more, in my opinion. It would be interesting, but it would not change our interpretation of the Chaco Phenomenon very much. My point is this--however the supernova pictograph is interpreted, the result is only of minimal value archaeologically. It simply does not constitute the kind of information that archaeologists rely on to address the questions of cultural process that are of primary interest to them. We already know that the Pueblo Indians watched the sun and were aware of what went on in the heavens. So how would the knowledge that the people of Chaco Canyon recorded a unique event, such as the supernova, help us? It certainly does not contribute much to the explanations of changes in human behavior that the archaeologists of today seek to find. Sad to say, something that the public and at least some archaeoastronomers find so fascinating simply does not contribute to the primary data base of the modern archaeologist. I hope that you are beginning to appreciate the nature of the differences between our two disciplines, as well as the magnitude of effort needed to integrate them.

Basically, these differences hinge on the distinction between the common and the unique, the mundane and the spectacular, the secular and the ceremonial. Astronomers look up, archaeologists look down. By their very nature, the data of the archaeoastronomers are unique and spectacular. The data of the archaeologists are, in contrast, more common and generally much less spectacular. This might be referred to as the "ho-hum" aspect of archaeology. Did you ever go to an archaeological convention and listen to some guy describe the interior dimensions of 34 bell-shaped storage pits? No wonder people yawn when they see us coming.

Again I digress. To return to the two kinds of data bases, the distinction would be less important were it not for the archaeologists' desire to go beyond description of past human behavior and to attempt to explain it. Explanation is sought in the context of some external factor, such as the physical and/or social environment, and the methodology of explanation frequently lies in the realm of quantitative analysis. Thus, archaeologists search for redundancy and patterns of behavior in order to seek and verify explanations of that behavior. The search includes analysis of covariation within the archaeological record--covariation that can be isolated and verified statistically.

This quest is quite different from that of the archaeoastronomer, who generally deals with the unique, not the redundant, and with the sacred, not the secular, thus making quantitative treatment of the data more difficult and explanation more challenging. To be quite candid, from the archaeological perspective, astronomical interpretations of the archaeological record frequently have been based more on simple correlation than on actual, verifiable covariation. Identification of covariation is difficult to accomplish in the realm of ritual, and this suggests why archaeologists tend to shy away from that realm in seeking explanation of the phenomena with which they deal.

But differences in primary data do not mean that archaeoastronomers should not attempt anthropological explanations, or that archaeologists should not consider the realm of ritual. I will deal with the latter in more detail when I discuss Chaco Canyon. With regard to the former, in my opinion astronomers need to go beyond the realm of description of the phenomena they observe, beyond the actual recording of these phenomena (which sometimes becomes an end in itself), and make more attempts to interpret the purpose and function of these phenomena in a larger social context. In brief, they need to explore the explanation of the archaeological sites they find in broader, more anthropological terms in order to contribute significantly to the interpretation of the archaeological record. When astronomers deal more with the cultural context of sites in an area, and when archaeologists deal more with the ritual aspects of the same sites, we will begin to see progress in the maturation of archaeoastronomy as a truly integrated discipline.

I would like to mention our research in Chaco Canyon as an example of how to approach the integration of the two disciplines discussed thus far. By "our," I mean the work that the Chaco Center, or Chaco Project, carried out for 10 years in the San Juan Basin, that 26,000 sq mi expanse of greasewood, saltbush, and strip mines lying to the northwest of Albuquerque. The Chaco Center was a National Park Service research organization located on the campus of the University of New Mexico that carried out archaeological and related research in and around Chaco Canyon between 1972 and 1985.

I mentioned that archaeologists shy away from ritual explanations of the phenomena they observe because such explanations are very hard to verify. For this reason, the ritual explanation is one that is resorted to only when other explanations are exhausted. Put differently, when they can't think of anything else to call it, archaeologists label it "ceremonial." Basically, that is what has happened in our research at Chaco, at least from one perspective.

To summarize our work there, the most parsimonious explanation or model, as we call it, of what we see there archaeologically is that the canyon functioned primarily as a ceremonial center, which received periodic pilgrimages from outlying areas, and that a great deal of real "business" involving day-to-day human survival in the San Juan Basin was carried out at those times under this ritual metaphor. The ritual, then, functioned as a means of gathering people together for the exchange of information and material goods. Note that, in our view, the true function of the ceremonial system was to regulate the exchange of goods and services throughout the basin.

Although we have arrived at a "ritual explanation" of an archaeological manifestation, it was arrived at neither lightly nor hastily. Only after other possibilities were examined, analyzed, and discarded as empirically invalid did we accept the ceremonial model as the most likely, and then only when the latent economic function could be assumed as an integral component. Thus, this is not a simple "We don't know what it is, thus it must be ceremonial" explanation of something as complex as the Chacoan system.

If this model is true, and right now it is just a hypothesis, then the challenge to archaeologists is to verify it. By the same token, in my opinion, the challenge to the field of archaeoastronomy is to aid in that process. This must be done by analyzing the ceremonial aspects of Chacoan sites in the broader cultural context of economically based alliance and exchange networks, which were basically an ecological adaptation to a marginal environment in the San Juan Basin. Such analysis is much easier to describe than to do, and it will involve close cooperation between archaeoastronomers and archaeologists interested in the Chaco Phenomenon. It will also involve, in my opinion, a change in the research perspective of archaeoastronomy in Chaco: a change from an emphasis on the descriptive aspects of the ceremonial sites to an emphasis on explanation of these sites relative to the larger cultural context.

Not long before this symposium took place my colleague Tom Windes and I were discussing how much we have learned about Chacoan archaeology since we started in 1972. It really is quite fascinating, since at the beginning we had virtually no idea that Chaco was the focus of a system permeating the entire San Juan Basin, no knowledge of the road network, or of the nature and timing of the climatic episodes, or of the importation of 200,000 trees to build the pueblos, no concept of pilgrimage festivals, and so on. We know these things now because we established a series of research objectives and systematically attempted to verify them empirically. The point is that scientific model building progresses through a series of theoretical/empirical trials.

Again, in my view, archaeoastronomy needs to follow roughly the same steps in progressing through its own series of testable models, in Chaco or elsewhere, in order to become more firmly integrated with, and beneficial to, the anthropological discipline from which it draws its primary data.

Going beyond Chaco for a moment, I would like to return to the question of why the public becomes so transfixed by archaeoastronomical discoveries, such as solstice markers. When such a marker is verified by astronomers through observation, the public seems to acquire an almost hypnotic fascination with it, crowding to the site on the appointed day to watch the thing perform (not unlike Old Faithful, actually). When you think about it, these performances are not much of a subject for awe and mystery. Consider, on the other hand, how mysterious it would be if the site did not perform on the solstice.

What puzzles me, I guess, is why laymen are so thrilled by their own personal discoveries that the Indians knew about solstices. Consider for a moment the fact that Native Americans lived very close to nature in this area for at least 12,000 years before the Caucasians arrived. Now, you can't live that close to a phenomenon for that long without observing something about it. Under the same circumstances, even Anglos would have figured out that the sun's position on the horizon varies with the seasons. Thus, the fact that the Anasazi tracked and marked the solstices should not amaze us, or even make us blink twice, yet it seems to do so. My hunch is that what we are witnessing here is a "rediscovery" of the sun's path by modern Anglo populations. Now that we have had an energy crisis, and we even get tax breaks for solar improvements, we as a culture have rediscovered the sun's movements and are beginning to track solstices again, as we undoubtedly did in our own European homelands centuries ago. This rediscovery is a function solely of our having been very effectively insulated from our environment for a long time; now, in the process of having to insulate our homes, we are finding out things about the real world we thought we never knew. Thus, in my view, the amazement we display over prehistoric solstice markers is largely a function of our own recently acquired ignorance about the environment.

This brings me to what I consider to be the most important challenge to the field of archaeoastronomy--the need to demonstrate to the public that there is more to all of this than the mechanics and aesthetics of solstice markers and that the achievements of the Anasazi were much more sophisticated than simply tracking the sun's movements along the horizon. We need to examine the behavior behind the markers to see how that behavior was integrated into the larger socioeconomic system that we now postulate as the key to survival in the San Juan Basin in the tenth and eleventh centuries. That socioeconomic system (that is, the Chaco system) was a truly significant achievement on the part of the Anasazi, one that may have been a virtually unique adaptation among cultures in similar environments on this continent. The potential contribution of archaeoastronomy to our understanding of that socioreligious system is both considerable and important, and it needs to be addressed systematically by anthropologically oriented astronomers.

Let me hasten to add that I am not the first person to recognize this challenge confronting archaeoastronomy, or even to articulate it. Far more knowledgeable people than I have written of this, two of the most recent being Dr. Anthony Aveni and Dr. Jonathan Reyman, both participants in this symposium. I would like to quote from Dr. Aveni's recent synthesis of archaeoastronomy:

> While we may demand that anthropologists and archaeologists learn astronomy [heaven forbid!, I would add], we have not advocated strongly enough the more difficult maxim that astronomers interested in archaeoastronomical research ought to become anthropologically oriented. . . . We must become more conscious about the relationships ancient astronomers sought between the heavens and agricultural cycles, meteorology, and environmental affairs in general. The . . . course followed by many investigators simply assumes "they were like us." It does not

> look enough at the needs and behavior patterns of ancient civilizations as they might relate to their astronomical systems [1981:S6-S7].

Amen, Dr. Aveni, and well said.

Although I began this discussion by emphasizing the disparity between archaeology and astronomy, I will end by stressing the commonality. Working together with a common understanding and a common approach to mutually agreed upon problems, archaeologists and astronomers can make significant contributions to the interpretation of past life systems. It will not be an easy task, but then remember that when the going gets rough . . . the archaeologists call in the graduate students.

As a keynote lecture, this certainly has been the most low-key aspect of this symposium on astronomy and ceremony in the prehistoric Southwest. Fortunately, the high-key parts, indeed the highlights, are yet to come. Over the next several days we will have the unique opportunity to hear a number of nationally recognized scholars address some of the issues that I have mentioned along with many others that will come to light. The true value of this symposium, of course, lies in the encouragement of communication among astronomers, prehistorians, and the public to facilitate the interplay that Aveni, Reyman, and others have called for as essential to the development and recognition of archaeoastronomy as a valuable discipline in its own right.

REFERENCES CITED

Aveni, Anthony F.
1981 Archaeoastronomy in the Maya Region: A Review of the Past Decade. Archaeoastronomy 3(JHA 12):S1-S16.

Chapter 2

Archaeoastronomy in the Southwestern United States:

A Neighbor's Eye View

Anthony F. Aveni

THE PSYCHOLOGY OF ARCHAEOASTRONOMY

There is no greater drag on the study of the perception of nature by cultures past than the methodological anchor that shackles us to the present. Bound as we are by our modern scientific approach to the natural world, it is all too tempting for us to make our distant predecessors behave as we do. We seek scientific explanations of the natural world through model building and testing; we strive to achieve the ultimate wedding of theory and fact by discovering and fine tuning certain laws of nature. To achieve these goals, we have greatly extended our senses in the past few centuries through the use of complex technical aids. Moreover, we have carefully developed and nurtured an objective logic of scientific reasoning that attempts to reduce observed phenomena to ever more refined phenomenal subsets--a reduction process that must necessarily proceed in a hierarchical pattern before nature's events can be reassembled into an understandable, connected whole. These are the objective desiderata that characterize science in the Western world. And our science has, at least since the industrial revolution, been part of a cultural milieu emphasizing the subjective notion of progress and the ultimate mastery of and control over nature.

It is not my intention either to evaluate or to pronounce moral judgment upon the conduct of modern science. I wish only to suggest that in both the popular and the professional literature there are instances in which we characterize ancient cultures, without justification, as having behaved uncannily like us, and that, as a result, we arrive at misleading, indeed demonstrably incorrect, conclusions about human behavior. If we desire to understand the astronomies of ancient cultures, we must be careful not to confuse our tools and methods, our aims and goals, and our outlook and cosmic view with theirs. I shall attempt to show that we must pay close attention to the assumptions we make about each culture whose astronomy we study. I contend that we shall arrive at the truth in our interdisciplinary studies only by attempting to view the universe through the eyes of the people we study. Generally, this means we must integrate rather than differentiate data from realms we often view as incompatible, e.g., astrology and astronomy or religion and metrology. While such an approach runs counter to our normal scientific procedure, I shall try to demonstrate that such a course is capable of producing results that converge with those derived from cultural anthropology--the real home ground of archaeoastronomy.

Let me begin with one of the grossest examples that I know in which a thick coat of presentist paint is applied to the past. This is an example that many of my colleagues might not deem even worthy of discussion, yet I believe that it is an important one because it has had a profound influence on the layperson's belief system regarding the past. In one of his many popular books on the great achievements of past civilizations, popular writer Erich von Daniken (1971:vii) states, "Incredible technical achievements existed in the past. There is a mass of know-how which we have only partially rediscovered today."

Von Daniken looks for an explanation of the past in present terms, because the things and ideas of the present are the only ones with which he is familiar. He does not bother to examine the cultures that produced the artifacts he discusses. Moreover, his view of ancient man's orientation to the natural world is that of one who does not observe nature for himself. But that covers most of us, for we live our lives in an artificially controlled environment. Our involvement with technology has freed us from the need to observe the sun in order to tell time or to witness the cycle of lunar phases to understand when we can expect faint natural light during the normal hours of darkness. Today we are all quite ignorant of the procedures and tools required to determine directly and accurately the length of the day, month, and year or to predict eclipses. When we learn, through their hieroglyphic inscriptions, that the Maya Indians of the Yucatan achieved with surprising precision all of these astronomical feats more than 1000 years ago, the interpretation that leaps most readily to mind is that they, like us, must have possessed technological abilities far in advance of those of most primitive cultures. Indeed, von Daniken, overwhelmed by what he finds, concludes that a technology must have existed that was advanced even beyond our own.

Given this tight circular chain of reasoning, von Daniken "discovers" that many elements of ancient cultures were very like our own. In one of the most popular examples, the ground drawings on the desert in Nazca, Peru, are interpreted to be runways upon which astronauts, visiting us from another world, landed their spacecraft. In another case, von Daniken describes certain odd face masks worn by Maya gods sculpted on stelae as the space helmets of these visiting astronauts. As he examines the data of the past, von Daniken molds the image of our ancient ancestors into one that looks more and more like a mirror image of ourselves. The denizens of the past become demystified when we find that they shared our own ideas and concerns, and we feel less uneasy about cohabiting the same planet with them. We are comforted to know that our ancestors were interested in doing the same things in the same way. A sense of persistence and continuity gives the common person a feeling that perhaps there is something good about today's version of the human race.

But surely this discussion of the uninformed fringe is hardly worth our time. Most educated readers dismiss von Daniken's reasoning as well as his conclusions. In a sense, however, a similar line of thought pervades some of the more scholarly studies of the achievements of ancient cultures.

Jacquetta Hawkes (1967:74), in her critique of the forays by modern astronomers into the discipline of archaeology, has stated that "Every age has the Stonehenge it deserves--or desires." Defending her humanistic turf, she railed against the then novel interpretation that Britain's most famous ancient monument was intended as an astronomical observatory and a computer for predicting eclipses. She complained that, at least in the scientists' minds, the standing stones were becoming simply parts of a technological device to record astronomical alignments to the sun and moon at horizon. The positions of the 56 Aubrey holes, chalk-filled depressions evenly spaced around the monument, were analyzed with a computer, and the results indicated that they could be components of a functioning "computer" used to count intervals between eclipses of the moon. As archaeologist R. J. C. Atkinson (1966) suggested, there can be no doubt the stones and holes could be used that way, but did that mean they were used that way? The astronomer's reply: "If I can see any alignment, general relationship or use for the various parts of Stonehenge, then these facts were also known to the builders" (Hawkins 1965:vii).

Employing precision surveying equipment, other workers in the 1960s extended the astronomical paradigm to a multitude of megalithic stone circles in the British Isles, suggesting that ancient earthworks were intended primarily as media for the expression of precise geometrical knowledge. Thus, Thom (1967b, 1971) posited that irregular-looking circles had been flattened deliberately here and there to form well-defined "egg-shaped" structures that incorporated both a precise measuring unit and a knowledge of Pythagorean geometry. According to this interpretation, the function of the circles lay well outside the socioeconomic and even the religious realm. Circle builders were thought to be desirous of purely reflective knowledge for the sake of knowledge, to be "competent engineer[s]," in the words of the modern engineer (Thom 1971:9). While the Stonehenge-observatory-geometry controversy is intellectually a far more serious debate than the Nazca-astronaut hypothesis, some of the underlying assumptions are the same. In each case, the explanation consists of reducing the unfamiliar to the familiar via the shortest route--by picking out the pieces of the foreign matrix that can be most directly and most easily interpreted from a contemporary viewpoint and ignoring the rest.

Why do I believe that many of these interpretations about the great works of ancient men are misleading? Because often they can be shown to be inconsistent with other facts we have in our possession. Even a little knowledge of Nazca art laid alongside pictures of the ground drawings reduces the runway argument to fantasy. I shall eschew the opportunity to debunk von Daniken and to offer alternative hypotheses for the explanation of the Nazca lines (this is the subject of a forthcoming study [Aveni in press]), but I will comment briefly on megalithic astronomy because this topic is relevant to the discussion of Chaco phenomena.

It is not a simple matter to formulate hypotheses concerning the megaliths that are based in our knowledge of the culture. The Stone Age Britons have left us no written record, few art works, and only a few archaeological remains. Before we can adopt the precise astronomical interpretation of Stonehenge, we need to establish that the claimed accuracy of the alignments is present in the standing stones, and then, if it is, we must explain why these people would share with us a penchant for calendrical precision. It is not unlikely that early hunter-gatherers, people who might convene for a time at a communal center such as Stonehenge to trade or worship, would desire to mark events in some sort of calendar. We might well expect them to track the sun and moon and, perhaps, the brightest planets. Alignments to the rising and setting points of astronomical objects would appear to be sensible constructs in their sphere of inquiry. But would they really care to measure the 9 arc-minute lunar wobble or a 2.720 $\pm$ 0.003 ft megalithic yard? (Nine minutes of arc is one-third of the width of the solar disk, and the 9 arc-minute wobble varies in a complex way over a period of nine months. In the opinion of this author, the relevant orientations at Stonehenge have not been established with satisfactory statistical significance; see also Ellegard 1981; Ruggles 1982, 1984.)

Only the assumption that the builders of Stonehenge were like us can save the day. We must be prepared to believe that, captivated by the finer details of lunar motion and possessing as much time in their day for reflection upon esoteric matters as we possess in our contracted 40-hour work week, these Stone Age people pursued knowledge for its own sake. It is a glorious and romantic notion to think of them as a race of prehistoric Einsteins, but it is an assumption that is wanting in supportive evidence. While the precision claimed for Stonehenge by a scientific investigator can be dismissed as a minor detail in the argument, the cultural implications of such precision cannot. The claim of precision was the catalyst that made Stonehenge such a center of controversy and is, in fact, the feature of the whole argument that arouses most readers' interest. We stand in awe when we view our ancestors in a new light--as cardboard images of ourselves in quest of the same fine detail. Often when dealing with the science of non-Western cultures of the past, we dwell on detail and precision only because these are the aspects of their fund of information that we can measure and interpret, but sometimes we carry this too far. We take the bits and pieces of ancient knowledge revealed in the cultural remains, carve and mold them, chip away at them, until we create an imagined system of use and acquisition of knowledge that is just like ours.

ASTRONOMICAL CONCEPTS AND STRATEGIES IN THE SOUTHWESTERN U.S.: THE OLD WORLD PARADIGM

So far I have suggested that in archaeoastronomical studies, Nazca lines and British megaliths are among those artifacts that have been most vulnerable to explanation by unsubstantiated hypotheses. Next I shall argue that many of the artifacts of the southwestern U.S., which I believe to be the next most vulnerable area, are often explained in terms of concepts that are borrowed uncritically from the study of the megaliths. Subsequently I will suggest that lessons can be learned about the nature and role of astronomy in ancient cultures by considering case studies in Meso- and South America, where we possess a greater corpus of cultural and historical data.

Table 2.1 presents data concerning several arbitrarily selected case studies in which astronomical phenomena were offered by the investigators as best fits for alignments. The table includes European megalithic sites (Group A), sites from both sedentary and mobile groups in western North America (Group B), and sites of the high cultures of Mesoamerica and South America (Group C), although grouping

Table 2.1. Astronomical alignments proposed for various sites

REFERENCE	SITE	EVENT: June Solstice	December Solstice	Equinoxes	Cardinals	Solar Zenith*	Solar Anti-zenith*	Other Solar	Lunar Stand-stills	Venus Stand-stills	Stars
Group A (European Megalithic)											
Hawkins (1963)	Stonehenge, Great Britain	+	+	-	-	-	-	-	+	-	-
Ponting and Ponting (1982)	Callanish, Great Britain	-	-	-	-	-	-	-	+	-	-
Thom (1967b, 1971)	various sites, Great Britain	-	-	-	-	-	-	-	+	-	+
Group B (Western North America)											
Autrey and Autrey (1979)	Zodiac Ridge, Arizona	+	+	+	-	-	-	-	+	-	+
Eddy (1974)	Medicine wheels, Wyoming and Canada	+	-	-	-	-	-	-	-	-	+
Evans and Hillman (1979)	Casa Grande, Arizona	+	+	+	-	-	-	-	+	-	-
Hudson and Underhay (1978)	Chumash sites, California	+	+	-	-	-	-	-	-	-	-
Reyman (1976)	Pueblo Bonito, New Mexico		+	-	-	-	-	-	-	-	-
Sofaer et al. (1982)	Fajada Butte, New Mexico	+	+	+	-	-	-	-	+	-	-
Williamson et al. (1977)	Casa Rinconada, New Mexico	+	-	-	+	-	-	-	-	-	-
Williamson et al. (1977)	Hovenweep, Utah	+	+	+	-	-	-	-	-	-	+
Williamson et al. (1977)	Pueblo Bonito, New Mexico	-	+	-	+	-	-	-	-	-	-
Williamson et al. (1977)	Wijiji, New Mexico	-	+	-	-	-	-	-	-	-	-
Group C (Meso- and South America)											
Aveni (1981)	Cuzco, Peru	+	+	-	-	+	+	-	-	-	+
Aveni et al. (1975)	El Caracol, Chichen Itza, Mexico	-	-	+	-	+	-	-	-	+	+
Aveni et al. (1975)	Uxmal, Mexico	-	-	-	-	-	-	-	-	+	-
Aveni and Gibbs (1976)	Teotihuacan, Mexico	-	-	+	-	-	-	-	-	-	+
Aveni (1977)	Copan, Honduras	-	-	-	-	+	-	-	-	-	-
Aveni and Gibbs (1976)	Templo Mayor, Tenochtitlan, Mexico	-	-	+	-	-	-	-	-	-	+
Aveni and Linsley (1972)	Monte Alban, Mexico	-	-	-	-	+	-	-	-	-	-
Carlson (1976)	Palenque, Mexico	-	+	-	-	-	-	-	-	-	-
Dearborn and White (1982)	Machu Picchu, Peru	+	-	-	-	-	-	-	-	-	-
Ricketson (1928)	Uaxactun, Guatemala	+	+	+	-	-	-	-	-	-	-
Rivard (1970)	El Castillo, Chichen Itza, Mexico	-	-	+	-	-	-	-	-	-	-
Schele (1976)	Palenque, Mexico	-	+	-	-	-	-	-	-	-	-
Tichy (1982)	various sites, Mexico	+	+	+	+	-	-	+	-	-	+
Zuidema (1981, 1982)	Cuzco, Peru	+	+	-	-	-	-	-	-	-	-

*Solar zenith and antizenith do not apply to Groups A and B

of so divergent an array of cultures as those in Groups B and C might be difficult to justify. Each plus (+) on the table represents an astronomical phenomenon that the authors of the cited works argue is marked or recorded at the site in question.

One can draw a number of interesting conclusions from scanning this table. The first is that nearly every ancient culture is claimed to have observed the sun at horizon at the solstices, and most of them at the equinoxes. This comes as no surprise since, where available, the ethnographic/ethnohistoric documentation for observation of these events is very strong. In many cases, and particularly in the Southwest, the method of observation is discussed explicitly (e.g., Parsons 1936).

A second clear pattern is that among the cultures of the tropics, observations of the zenith-antizenith sun at horizon often supplement or even replace other solar observations in the calendar. (The term <u>zenith-antizenith</u> refers to the two days when the sun passes the overhead position and the two days opposite that time in the calendar year. The observation applies only to the tropical latitudes, the only region on the earth where the sun can be viewed directly overhead.) Further study reveals that these solar stations are temporally more in tune with certain important agrarian phenomena, such as the start of the rainy season and/or the commencement of planting (Zuidema 1981), than are the solstices or equinoxes. Thus, the relative importance of this phenomenon probably has little to do with the degree of sophistication of the calendars of any of these cultures.

The table also shows that, except for a few passing mentions in Thom (1967b), stars are cited as reference objects in the New World but not in the Old. In practically every case the stellar referent bears a functional relationship either to another astronomical phenomenon (e.g., the heliacal rising of a set of stars occurring at intervals of a synodic month [as measured by the lunar phase cycle; Eddy 1974]) or to another, nonastronomical natural event (e.g, the heliacal rise of the Pleiades signaling the start of the agricultural year [Zuidema 1982]). Observations of multiple related phenomena could be taken to imply the development of a more complex calendar. The percentage of stellar references in each of the groups in the table is consistent with this suggestion, for as we proceed from Group A to Group C, we move in the direction of cultures known to have developed more complex calendars. On the other hand, the greater frequency of stellar references among the cultures represented in Groups B and C may simply be a result of the latter materials having been studied more thoroughly.

The final observation that I would like to make about the data in Table 2.1 is that lunar standstills are proposed as best fits to alignment data only for the sites in Groups A and B. Since these are among the most esoteric of the proposed astronomical observations, and since knowledge of the standstills has been tied to eclipse prediction, it would be worthwhile to review exactly what is implied when one suggests that the lunar standstills were observed.

The period of regression of the lunar nodes (intersection points of the moon's orbit with the ecliptic) is 6797.1 days

= 18.61 tropical years (years measured by the seasons)

= 230.17 lunar synodic months

= 249.78 lunar draconitic months (months measured by the interval between successive passages of the moon by a node of its orbit)

In relation to horizon observations, this period returns the moon to the same extreme point of the horizon. But because this period is not commensurate with the synodic and draconitic months, eclipses of the full moon need not and probably will not occur at the same time in successive cycles.

Often the lunar standstill period is confused with the cycle of Meton: 6940 days

= 19.00 tropical years
= 235.01 lunar synodic months
= 255.03 lunar draconitic months

This cycle marks the periodicity with which a given phase of the moon will occur on the same date of the year. Since the Metonic cycle, unlike the lunar standstill period, is an integral multiple of both the draconitic and synodic lunar intervals, a lunar eclipse is likely to occur on the same date as an eclipse during the previous cycle.

What is the source of the belief that the moon's extreme rise-set points, the so-called lunar standstills, were observed by ancient people? The earliest reference appears in a 1912 paper by Somerville. [Note: I am indebted to Robert Merritt and Alexander Thom who, in recent correspondence, directed me to a note by Thom in the commentary section of Antiquity (Thom 1967a:92) in which Thom credits Somerville (1912) with drawing attention "to a possible lunar line at Callanish." Hawkins (1965:185) also mentions this reference.] In a later paper, Somerville (1928:33-34) rejected any orientation hypothesis based upon lunar alignments because, as he put it,

> the necessary observations of the Moon--could not have been by its rising (or settings) because, except on the day of full moon either the rising of the Moon, or its setting, or both, take place during daylight, and so are not visible. Orientation consequently cannot probably be connected with moonrises, except, perhaps on certain full moon days.

Hawkins (1963, 1964, 1965) offered little explanation or justification when he included the lunar horizon limits in the Stonehenge alignment tests along with the solar extrema. He appears to have adopted them as a logical extrapolation of the solar horizon limits. When he achieved a positive alignment correlation and began to analyze the Aubrey holes as a seasonal eclipse counter, he then proposed the eclipse-observatory hypothesis for Stonehenge.

Hawkins (1964) also considered it possible that Diodorus's statement about a temple on the island of the Hyperboreans may have referred to Stonehenge and that the 19-year period mentioned in the statement could be connected with the lunar extrema and the nodal regression period. The quotation from Diodorus begins,

> They say also that the moon, as viewed from this island, appears to be but a little distance from the earth and to have upon it prominences, like those of the earth, which are visible to the eye. The account is also given that the god visits the island every nineteen years (1935:47).

In Hawkins's original publication in Nature (1963), the next sentence in the Diodorus passage was omitted. This sentence, which was included in Hawkins's book (1964:130), clearly identifies the 19-year period with the Metonic cycle and not the nodal regression cycle:

> and for this reason the nineteen-year period is called by the Greeks the "year of Meton."

Thom (1971:11) justifies the claim for observations of the lunar extrema by inferring that "men living on the coasts of the Western Ocean must have noticed the connection between the tides and the Moon. Probably the connection was one of the factors that caused Megalithic men to study the movements of the Moon." According to Thom, the study was carried further until ancient astronomers were "able to predict which new or full Moon would give rise to an eclipse of the Moon or the sun." One reason often given for the desire to predict eclipses is access to specialized knowledge resulting in power; however, there are no ethnographically documented examples of the use of knowledge of eclipse predictions as a source of power.

Thom also suggests that "The importance to Megalithic people of the moon for illumination in the winter and for tidal prediction evidently caused them to make a study of its movements" (1971:23). Based on this assumption he concludes that "The 18.6-year cycle would obtrude itself. . . . Analogy with a probably earlier study of the solstices would suggest the use of a horizon mark with a back-sight" (Thom 1971:23).

These assumptions, taken together with the speculation that sooner or later someone would notice that "eclipses happened at the new or full Moon nearest the date when the moon attained its absolute extreme positions" (1971:23), are offered to account for the alleged discovery of the 9-minute lunar wobble.

The desire to predict seasonal eclipses would appear to be the only raison d'être for paying attention to the lunar standstills. After explaining how the changing positions of new and full moons at horizon could be used to mark the 18.6-year cycle, Thom (1967b:21) suggests that ancient man must have been challenged by eclipses. It was an "impressive spectacle," and his "desire to master eclipse prediction" motivated him further. Now, since <u>we</u> know that eclipses can happen only when the moon is at one of the nodes of its orbit, we are asked to believe that ancient man observed that "no eclipses occurred near the solstices when the full moon was in either of its extreme positions but only in the years which lay midway between these" (Thom 1967b:21).

Stated in pure and simple terms, both Thom and Hawkins are referring to the empirically derivable statement that eclipses occur at progressively earlier times in the year, during the so-called eclipse seasons. The eclipse year, by which all eclipses may be timed, is shorter than the seasonal year and regresses through the seasonal calendar, completing a full passage through it in the course of 18.6 years. <u>In our culture</u> we connect this statement with another fact, also observationally derived, that the variation of the altitude of the moon on the meridian and of the limits of the rise-set positions of the moon at horizon follow the same cycle. But to assume that other cultures would discover and connect these same cycles or regard such a connection as important or useful constitutes the narrowest form of "astro-ethnocentrism."

I have elaborated on these arguments in some detail simply because, as far as I can determine, all claims that prehistoric people marked the lunar standstills ultimately rest upon this kind of ethnocentric reasoning. Too many authors (myself included [Aveni 1972]) have uncritically included the lunar standstills in lists of phenomena likely to have been observed by ancient people. There is no evidence that people living in the southwestern U.S. either desired to predict or were capable of predicting eclipses. There is, on the other hand, very persuasive evidence that the Maya (included in Group C in Table 2.1) did predict eclipses with great accuracy (Thompson 1972), yet there is not a single acceptable alignment with the lunar standstills in all of the known Maya architecture. Why? Because the standstills offer little useful data bearing on the prediction of eclipses or indeed on the prediction of anything. Furthermore, there is an easier way to predict eclipses, and we have evidence that both the Maya in the New World and the Babylonians in the Old forecast their occurrence with remarkable accuracy by grouping sequences of full moons in groups of six, with an occasional group of five inserted at the appropriate place in the count. By this means they were able to recognize, in modern astronomical parlance, the near equivalence between the eclipse half-year and the six-lunar-synodic-month period:

Eclipse half-year = 173.32 days

Six lunar synodic months = 177.18 days

Five lunar synodic months = 147.65 days

It might be argued that quite aside from their relationship to eclipse predictions, the lunar standstills are easy to observe, and furthermore (as was suggested above) that they are an obvious extension of the solar standstill phenomena. But does this reasoning justify the major effort that some field investigators have put into the search for lunar standstill alignments?

I believe that the lunar standstill phenomenon has received so much attention in studies of archaeoastronomy because it is an artifact of western European astronomy that has become firmly engrained in our minds. In spatial terms we think of the standstill cycle as the time required for the line of intersection between the lunar orbit and the ecliptic (i.e., the line of nodes) to make a full circuit around the ecliptic. In astronomies that do not utilize the concepts of orbit or node the standstill phenomenon may well have played no role at all, and until convincing evidence well beyond the level of coincidence is presented, evidence that can be documented from studies that are not connected with the alignments, I believe one ought to reserve judgment on the question of whether lunar standstills were important in ancient astronomy.

Even if Anasazi or British megalithic alignments are correlated with the lunar standstills (and there has been much criticism of Thom's work on the latter [e.g., Ruggles 1984]), there is no convincing astronomical reason for ancient people to incorporate such orientations--even if they are only approximate orientations (Burl 1982)--into their architecture or environment. In the light of recent developments, I believe it is time for a critical reassessment of the lunar-standstill orientation hypothesis both in the Old and in the New World.

A related issue that might be raised here concerns proposed orientations to the Venus standstills in Mesoamerican sites (e.g., Aveni et al. 1975). One might ask, if lunar standstill alignments cannot be supported, why suggest alignments with Venus standstills? The answer is that the written records that survived the Conquest not only are consistent with suggestions that such observations were made, they actually indicate the possibility of such observations. The arrangement of the Venus Table in the Dresden Codex strongly implies Mayan knowledge of an 8-year Venus cycle, a cycle that is manifested through the standstills. Furthermore, observations of the disappearance and reappearance of the planet at horizon, the two most important events depicted in the written calendar, result in a critical interval of absence of the planet from the sky that exhibits delicate seasonal variations (Aveni 1982) and that was related to astrological prediction (Thompson 1972).

A FUTURE COURSE OF ACTION

If we wish to identify astronomical phenomena that are likely to have been of interest to prehistoric people in the southwestern U.S., where should we look? I would suggest that the underutilized historical literature bearing on the practice of astronomy in the Southwest may prove helpful. Young and Williamson (1979) offer a number of Zuni origin myths with celestial referents. McCluskey (1979, 1982), in his analysis of the Hopi lunar count, refers to a host of observations from Stephen (Parsons 1936) and other early ethnographers concerning the location of the sun at horizon on dates of agricultural and ritual significance (e.g., planting of sweet corn, planting of the principal corn crop, dating of the Snake ceremony or of the Niman Kachina). Searching out solar orientations for these dates would constitute a far more imaginative and likely more fruitful strategy than simply looking at the solstices or, still worse, halving or quartering the intersolstice periods in order to divide the year into equal parts (Somerville 1928:33).

McCluskey's research on Hopi astronomy implies that corrections of the calendar may have necessitated a shifting of solar events observed along with the related solar horizon positions. Such alterations pose something of a problem for the archaeoastronomer who, in McCluskey's terms, might tend to "look at the astronomical systems of traditional cultures in a timeless, ahistorical manner, assuming them to be static and unchanging" (1979: 178). McCluskey's results on the lunar calendar, which seem to converge with those of Bunzel (1932) and Ellis (1975), suggest to me that the lunar cycle of interest may well have been the Metonic cycle (or something like it) rather than the nodal regression period. If we think that the relation of the synodic lunar and solar calendars was a problem for them, then it should be a problem for us.

Finally, I suggest that the planetary extremes might be worth looking for in Southwestern archaeoastronomy, especially since Young and Williamson (1979) have pointed out that a number of old ethnographic sources refer to the observation of the morning star at winter solstice. Though one difficulty in working with the planetary extremes is that they vary over relatively short intervals of time, with the advent of more accurate archaeological dating methods it may be possible to draw limited conclusions about such observations.

In sum, as Reyman (1975) has already suggested, we need to make better use of the ethnohistoric and ethnographic data as starting points for our hypotheses rather than trying to find data to support the alignments that we propose on the basis of analysis of the archaeological remains.

My analysis of Table 2.1 indicates that the research strategy of the investigators working in the Southwest seems less like that of the archaeoastronomers working in the southern regions of the New World and rather more like that of the Old World archaeoastronomers, who expend a great deal of energy searching for alignments. As Hoskin (1983:10-11) puts it,

> In the New World, it is possible, even necessary to consider written evidence and oral traditions in addition to examining the archaeological remains, and the approach is inevitably interdisciplinary. By contrast, the evidence from megalithic culture consists very largely of stone remains, whose purpose has to be inferred from archaeological and astronomical considerations. As the stones have often been disturbed, statistical treatment is required. Many archaeologists find the astronomical and statistical techniques difficult of access, while the mathematicians and astronomers sometimes blunder in matters of archaeology. Yet in both cases, because of the paucity of the evidence, they are forced to adopt a more rigorous stance than many of their New World counterparts.

The lesson we have learned from the study of the practice of astronomy among the cultures south of our borders is that they were far more imaginative than we had anticipated and that their astronomy was very different from our own. By combining alignment study with ethnohistorical research we have discovered some real surprises in Mesoamerican and Andean ceremonial architecture.

1. The principal temple of ancient Mexico City (Tenochtitlan) aligns 7 degrees 6 minutes S of true E and yet points to the equinox sunrise (Aveni and Gibbs 1976).

2. The primary orientation to the western horizon of Cuzco, 13 degrees 30 minutes N of W, indicates the place where the sun sets on that day in the year that is the reciprocal of the day when it passes the zenith (Aveni 1981; Zuidema 1981). This date also marks the official commencement of the planting season in the middle elevations in the valleys of Cuzco.

3. Among the principal astronomical directions of significance in certain contemporary communities of the Andes is the set consisting of the rising position of the Pleiades and its opposite, the setting position of the stars of Scorpio (Urton 1982).

Since none of these conclusions about indigenous American astronomy could have been reached on the basis of the archaeological or astronomical evidence alone, and since one property that Groups B and C of Table 2.1 have in common is that an ethnohistoric record exists, I offer a bit of neighborly advice to all researchers involved in North American archaeoastronomy that by now should be apparent. Drop the methodological paradigm of "alignment hunting" that is characteristic of Old World archaeoastronomy and apply with greater fervor the interdisciplinary "guided search" technique that has proven more fruitful in Meso- and South American studies.

ACKNOWLEDGMENTS

I am grateful to the editorial staff for their helpful comments.

REFERENCES CITED

Atkinson, R. J. C.
1966 Moonshine on Stonehenge. Antiquity 40:212-216.

Autrey, N., and W. Autrey
1979 Zodiac Ridge. In Archaeoastronomy in the Americas, edited by R. Williamson, pp. 81-100. Ballena Press, Los Altos, California.

Aveni, A. F.
1972 Astronomical Tables Intended for Use in Astro-Archaeological Studies. American Antiquity 37:531-540.
1977 Concepts of Positional Astronomy Employed in Ancient Mesoamerican Architecture. In Native American Astronomy, edited by A. F. Aveni, pp. 3-20. University of Texas Press, Austin.
1981 Horizon Astronomy in Incaic Cuzco. In Archaeoastronomy in the Americas, edited by R. Williamson, pp. 305-318. Ballena Press, Los Altos, California.
1982 The Moon and the Venus Table: An Example of Commensuration in the Mayan Calendar. Paper presented at the Conference on Ethnoastronomy, Washington, D. C.

Aveni, A. F., editor
in press The Lines of Nazca. University of Oklahoma Press, Norman.

Aveni, A. F, and S. Gibbs
1976 On the Orientation of Pre-Columbian Buildings in Central Mexico. American Antiquity 41:510-517.

Aveni, A. F., S. Gibbs, and H. Hartung
1975 The Caracol Tower at Chichen Itza: An Ancient Astronomical Observatory? Science 188:977-985.

Aveni, A. F., and R. Linsley
1972 Mound J, Monte Alban: Possible Astronomical Orientation. American Antiquity 37:528-531.

Bunzel, R.
1932 Introduction to Zuni Ceremonialism. Bureau of American Ethnology Annual Report 47:467-544.

Burl, A.
1982 Pi in the Sky. In Archaeoastronomy in the Old World, edited by D. C. Heggie, pp. 141-170. Cambridge University Press, Cambridge, England.

Carlson, J.
1976 Astronomical Investigations and Site Orientation at Palenque. In The Art, Iconography and Dynastic History of Palenque, Part III, edited by M. Greene Robertson, pp. 107-122. R. L. Stevenson School, Pebble Beach, California.

von Daniken, E.
1971 Chariots of the Gods. Bantam, New York.

Dearborn, D. R., and R. White
1982 Archaeoastronomy at Machu Picchu. In Ethnoastronomy and Archaeoastronomy in the American Tropics, edited by A. Aveni and G. Urton, pp. 249-260. Annals of the New York Academy of Sciences, Vol. 385. New York.

Diodorus of Sicily
1935 Book II. Harvard University Press, Cambridge, Massachusetts.

Eddy, J.
1974 Astronomical Alignment of the Big Horn Medicine Wheel. Science 184:1035-1043.

Ellegard, A.
1981 Stone Age Science in Britain. Current Anthropology 22:99-125.

Ellis, F. H.
1975 A Thousand Years of the Pueblo Sun-Moon-Star Calendar. In Archaeoastronomy in Pre-Columbian America, edited by A. F. Aveni, pp. 59-87. University of Texas Press, Austin.

Evans, J., and H. Hillman
1979 Documentation of Some Lunar and Solar Events at Casa Grande. In Archaeoastronomy in the Americas, edited by R. Williamson, pp. 133-136. Ballena Press, Los Altos, California.

Hawkes, J.
1967 God in the Machine. Antiquity 41:174-180.

Hawkins, G.
1963 Stonehenge Decoded. Nature 200:306-308.
1964 Stonehenge: A Neolithic Computer. Nature 202:1258-1261.
1965 Stonehenge Decoded. Delta Dell, New York.

Hoskin, M.
1983 History of Astronomy. IAU Information Circular No. 32, Commission 41. Institute for de Eksakte Videnskabers Historie, Aarhus, Denmark.

Hudson, T., and E. Underhay
1978 Crystals in the Sky: An Intellectual Odyssey Involving Chumash Astronomy, Cosmology and Rock Art. Ballena Press, Santa Barbara, California.

McCluskey, S.
1979 Transformations of the Hopi Calendar. In Archaeoastronomy in the Americas, edited by R. Williamson, pp. 173-182. Ballena Press, Los Altos, California.
1982 Historical Archaeoastronomy: The Hopi Example. In Archaeoastronomy in the New World, edited by A. Aveni, pp. 31-58. Cambridge University Press, Cambridge, England.

Parsons, E. C., editor
1936 Hopi Journal of A. M. Stephen, 2 vols. Columbia University Contributions to Anthropology 23. New York.

Ponting, M., and A. Ponting
1982 Decoding the Callanish Complex: A Progress Report. In _Archaeoastronomy in the Old World_, edited by D. C. Heggie, pp. 191-204. Cambridge University Press, Cambridge, England.

Reyman, J.
1975 The Nature and Nurture of Archaeoastronomical Studies. In _Archaeoastronomy in Pre-Columbian America_, edited by A. F. Aveni, pp. 205-216. University of Texas Press, Austin.
1976 Astronomy, Architecture, and Adaptation at Pueblo Bonito. _Science_ 193:957-962.

Ricketson, O.
1928 Astronomical Observations in the Maya Area. _Geographic Review_ 18:215-225.

Rivard, J.
1970 A Hierophany at Chichen Itza. _Katunob_ 7:51-55.

Ruggles, C.
1982 Megalithic Astronomical Sightlines: Current Reassessment and Future Directions. In _Archaeoastronomy in the Old World_, edited by D. C. Heggie, pp. 83-106. Cambridge University Press, Cambridge, England.
1984 Megalithic Astronomy: The Last Five Years. _Vistas in Astronomy_ 27:231-289.

Schele, L.
1976 Accession Iconography of Chan Bahlum in the Group of the Cross at Palenque. In _The Art, Iconography and Dynastic History of Palenque_, Part III, edited by M. Greene Robertson, pp. 9-34. R. L. Stevenson School, Pebble Beach, California.

Sofaer, A., R. M. Sinclair, and L. E. Doggett
1982 Lunar Markings at Fajada Butte, Chaco Canyon, New Mexico. In _Archaeoastronomy in the New World_, edited by A. Aveni, pp. 169-182. Cambridge University Press, Cambridge, England.

Somerville, B.
1912 Astronomical Indications in the Megalithic Monument at Callanish. _Journal of the British Astronomical Association_ 23:83.
1928 Orientation. _Antiquity_ 1:31-41.

Thom, A.
1967a Hoyle on Stonehenge: Some Comments. _Antiquity_ 41:95-96.
1967b _Megalithic Sites in Britain._ Oxford University Press, Clarendon.
1971 _Megalithic Lunar Observatories._ Oxford University Press, Clarendon.

Thompson, J. E. S.
1972 _A Commentary on the Dresden Codex, a Maya Hieroglyphic Book._ Memoirs of the American Philosophical Society No. 93. Philadelphia.

Tichy, F.
1982 Der Festkalendar Sahagun's. Ein Echter Sonnenkalendar? Lateinamerica Studien 6:115-137.

Urton, G.
1982 At the Crossroads of the Earth and the Sky. University of Texas Press, Austin.

Williamson, R., H. Fisher, and D. O'Flynn
1977 Anasazi Solar Observatories. In Native American Astronomy, edited by A. F. Aveni, pp. 203-218. University of Texas Press, Austin.

Williamson, R. H. Fisher, A. Williamson, and D. O'Flynn
1975 The Astronomical Record in Chaco Canyon, New Mexico. In Archaeoastronomy in Pre-Columbian America, edited by A. F. Aveni, pp. 33-43. University of Texas Press, Austin.

Young, M. J., and R. Williamson
1979 Ethnoastronomy: The Zuni Case. In Archaeoastronomy in the Americas, edited by R. Williamson, pp. 183-192. Ballena Press, Los Altos, California.

Zuidema, R. T.
1981 Inca Observations of the Solar and Lunar Passages Through the Zenith and Antizenith at Cuzco. In Archaeoastronomy in the Americas, edited by R. Williamson, pp. 319-342. Ballena Press, Los Altos, California.
1982 Catchillay: The Role of the Pleiades and of the Southern Cross and A and B Centauri in the Calendar of the Incas. In Ethnoastronomy and Archaeoastronomy in the American Tropics, edited by A. Aveni and G. Urton, pp. 203-230. Annals of the New York Academy of Sciences, Vol. 385. New York.

Chapter 3

Anticipation in Ceremony:

The Readiness is All

Michael Zeilik

HISTORICAL PUEBLO ASTRONOMY

Astronomy is integral to the ritual life of the historical Pueblos, both in symbol and in practice. The practical aspects include both sun and moon watching to ascertain correct times for planting and to initiate essential ceremonies. Although the historical Pueblos have suffered severe pressures from the Spanish and Anglo cultures, their traditions remain remarkably intact. Arguing by analogy, I assume that this astronomical tradition retains in its general character holdovers from prehistoric times (Ellis 1975) and the practices of the Anasazi.

This argument by analogy has some obvious weaknesses because there is historical evidence of changes in the Puebloan culture. On the other hand, Pueblo culture emphasizes the "old," ancestral ways as being of great value and considers that which is new to be dangerous. Despite the weaknesses, though, I consider argument by analogy in the context of the Puebloan culture to be a better way to work toward an understanding of the Anasazi than argument from the context of Anglo culture.

In order to achieve a complete understanding of prehistoric astronomy, we would have to evaluate it in its own cultural context, from the perspective of the Anasazi equivalent of a sun priest. That is, of course, impossible, but we can try to achieve the perspective of a Pueblo sun priest, the cultural heir of Anasazi astronomy, by means of a careful study of the ethnoastronomy of the historical Pueblos. By doing so, we can hope to avoid imposing the astronomy of the Western tradition upon that of the Anasazi and can come as close as possible to the frame of mind of the people of the past.

Although the art of sun and moon watching is mentioned frequently in accounts of the historical Pueblos, only Reyman (1971:114-122) has attempted to survey the literature for a large sample of Pueblos to infer the general trends--those that have the strongest roots in tradition and so are the most likely to have been carried over from the past. In order to develop a data base for ethnoastronomy I have examined closely the ethnographic studies of the historical Pueblos (Zeilik 1983b). So far I have ethnographic data on 19 pueblos (if the Hopi villages are considered separately, as I think they should). I have found no information so far on Tesuque, Pojaque, Sandia, Ysleta del Sur (Tigua), and Pecos that relates to astronomical practices.

As an astronomer, I have focused on the astronomical practices rather than those of ritual. In particular, I have searched for information on who does the observing, where the observations take place, how the observations are made, and when they are made. The results for the general characteristics of Puebloan sun watching to be discussed here are summarized below; details of this investigation are presented in Zeilik (1983b).

1. A religious office (sun priest) is invested with the responsibility for watching the sun.

2. Observations take place at sunrise (usually) and sunset (less often) from within or close to the pueblo; the place is rarely marked.

3. Horizon markers are most commonly used to establish the sun's seasonal position; light and shadow (through windows) is not as frequent as an observing method.

4. The most important times of the ritual calendar are the solstices, especially the winter; the sun priest also keeps track of the basic planting calendar from April to June.

5. The sun priest must be able to announce the solstices ahead of time; he does so by anticipatory observations made about two weeks before the solstice when the sun is still moving a noticeable amount along the horizon every day.

6. The site and technique for the anticipatory observations are usually the same as for the confirmatory ones.

One critical aspect of sun/moon watching is this: the sun priest must be able to anticipate ceremonial events. Pueblo culture requires a preparation time of at least four days for a ceremony so that the participants will be in the proper frame of mind to make it effective. Preparations may include the manufacture of prayer sticks, fasting, abstinence from sexual relations, and dance and song practice. This all takes time, which is why the sun priest at Walpi (a Hopi village on First Mesa), for example, announces Soyal (the winter solstice ceremony) publicly on December 10 or 11. The anticipatory observations typically begin on December 6 (Parsons 1936:5-6).

Anticipation solves an astronomical problem, especially for the solstices. From solstice to solstice the sun swings through an angle of some 60 degrees along the horizon in the latitudes that include the Southwest. The daily rate at which the sun travels this horizon arc varies during the year (Figure 3.1). At the solstices, the sun has no noticeable motion for four days according to the Hopi at First Mesa--a statement that is both ritually and astronomically correct. The sun "stands still," the meaning of the word solstice. At the equinoxes, the rising and setting sun moves from point to point on the horizon at the fastest rate--almost the sun's own angular diameter each day.

The astronomical problem for the solstices is this: because the sun is essentially not moving, it is difficult to tell by naked-eye observations at the time of the solstices which day is, in fact, the solstice. The best way, astronomically, to determine the solstice is

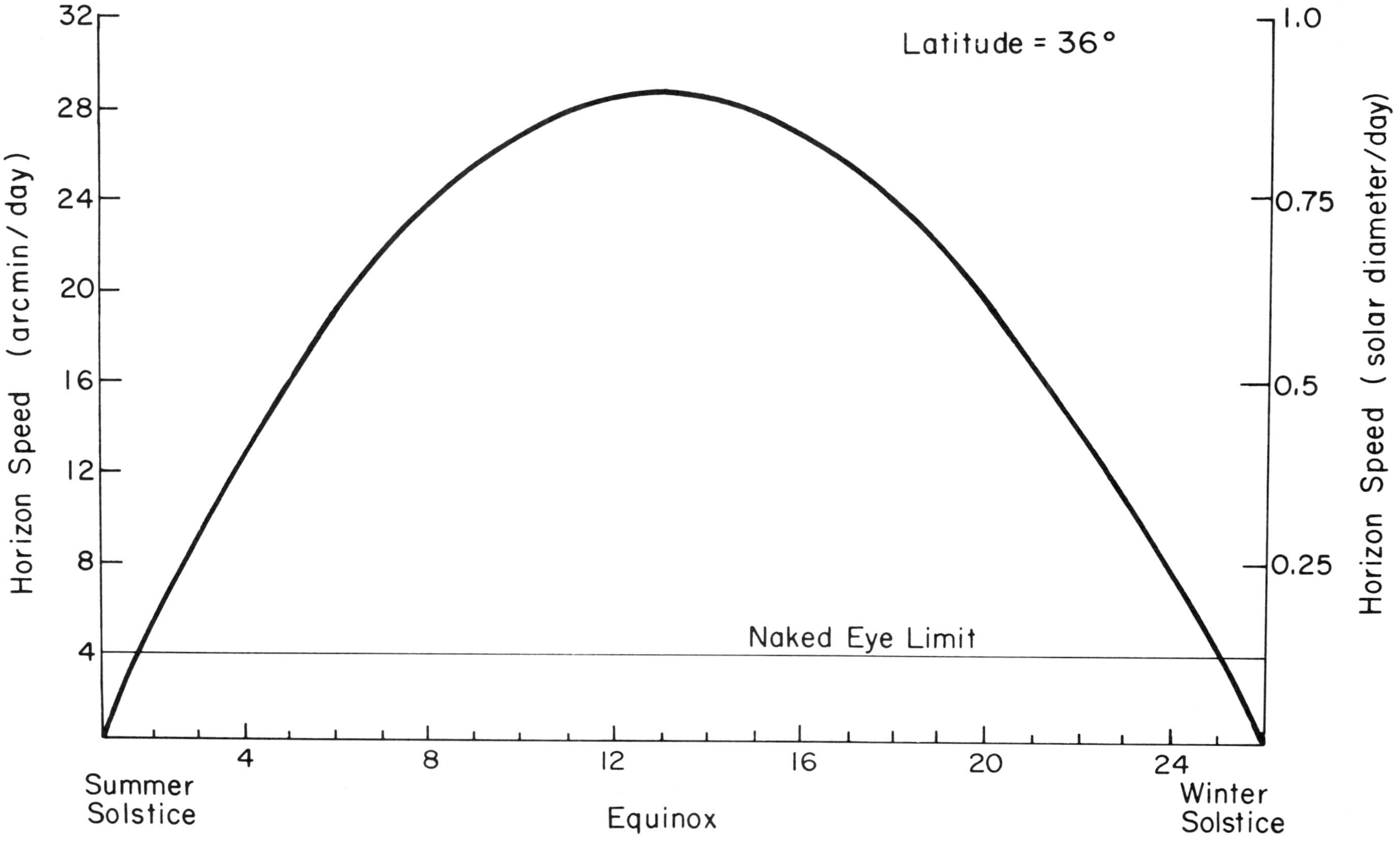

Figure 3.1. The angular speed of the sun along the horizon for one-half year for the latitude of the Southwest. The calculation assumes a flat horizon and no refraction.

to observe the sun before the solstice, while it is still moving observable amounts along the horizon. Using just the eye and horizon features, I estimate (based on my own sun-watching experience) that the minimum detectable solar motion is 4 arc minutes in a day under the best conditions ("naked eye limit" on Figure 3.1); double this number may be the limit reachable in practice. This rate of motion occurs roughly a week before the solstice. Observations made two weeks before solstice (or earlier) make the prediction even more reliable. The sunwatcher can then do a day count (perhaps using notches on a stick to keep a tally [Forde 1934:227]) to predict the day on which the solstice will occur. Of course, a few years of sun watching are required to establish the anticipation times for various horizon markers. But once established, the horizon calendar remains fixed for a long time (because it is not affected by precession), and the knowledge can be handed down to others.

That knowledge--the predictive dimension of priestly power (Reyman 1980)--is the esoteric knowledge of the sun priest. I judge that everyone at the pueblo knows when the solstice is occurring--that is public knowledge (Dutton and Marmon 1936:5). But only the sun priest and a few others know about the anticipatory observations. The difficulty of acquiring that knowledge can be inferred from Benedict's (1935:67) translation of the account of "The Sun's Son" of Zuni; it took the pekwin, the Zuni sun priest, eight years "to time the turning of the sun exactly." Even if this "eight years" is not meant literally, it is twice the sacred number four and so may convey a sense of the ritual importance of the training.

What is the meaning of "exactly" in the context of the Puebloan culture? In the Zuni tale, the pekwin is first 13, then 20, then 2 days off in his predictions of the actual day, and finally he predicts it "exactly." This statement implies that even two days difference is too much--that the pekwin was supposed to be able to predict the solstice to within one day. For other indications, I have surveyed the historical reports of the celebration times of the winter and summer solstices for Hopi and Zuni and compared them with the astronomical dates (Meeus 1983). For 15 records the mean difference between actual and celebratory dates is 1.6 days. If I exclude two cases when the difference was the worst (5 days), the average falls to 1.2 days. So in the Puebloan context, precision for the solstice prediction seems to be one day in principle and two days in practice. A note by Parsons (1925:120, note 184) confirms this inference with a specific example. She states that the sunwatcher at Hopi-Tewa on First Mesa from 1921 to 1924 announced solstice dates that "varied only a day or two from year to year."

One other key point is that the sunwatcher makes the anticipatory observations and the confirmatory ones from the same location using the same techniques, so the sunwatching station plays essentially a dual role.

Direct inference from historical ethnoastronomy has specific implications for archaeoastronomy, which are summarized below. Because anticipatory observations play such a crucial part in ritual, I suggest that in order for a prehistoric site to be judged as having an intentional calendrical function it should lend itself to both confirmatory and anticipatory observations (Zeilik 1983a). If it only confirms, I would say that the case for its intentional use is much weakened. I draw this conclusion from the fact that calendrical sun-watching stations in historical times serve both functions.

1. Potential sites will use horizon markers or the manipulation of light and shadow in buildings.

2. Observation sites with calendrical functions are most likely to be within or close to the pueblo.

3. Observation times are most likely at sunrise and less likely at sunset.

4. Horizon markers can be rather small changes in relief; shrines may be located at the sun's key positions on the horizon.

5. Winter solstice is the most likely event to be marked by a confirmatory alignment, and summer solstice next likely; a planting calendar (April-June) may also be marked.

6. The site and technique for the confirmatory observations should also work for anticipatory ones about two weeks before the solstices.

ANTICIPATION IN ANASAZI ASTRONOMY

I will apply the concept of anticipation of astronomical events to six Anasazi sites: three in Chaco Canyon (Casa Rinconada, Pueblo Bonito, and the three-slab site on Fajada Butte) and three at Hovenweep (Hovenweep Castle, Unit Type House, and Cajon Ruins). These sites all incorporate possible light and shadow patterns that may have been used for calendrical sun watching. Only the three-slab site does not incorporate a building. Among the historical Pueblos, a definite minority use light and shadow patterns rather than or in addition to horizon observations. One of these cases, the use of sunlight around noon at Isleta, does not serve a calendrical function, only a ritual one (Parsons 1932:290-293). Light and shadow techniques are historically integrated into buildings. Prehistoric analogs to these structures, because they are substantial, tend to survive much better in the archaeological record than the more ephemeral stations used for observing a horizon calendar. In fact, I expect that few horizon-watching stations would be detectable after a thousand years.

Casa Rinconada

Williamson and others (Williamson 1981; Williamson et al. 1975, 1977) have proposed that the northeast opening in the great kiva of Casa Rinconada served to confirm the summer solstice by channeling a beam of light into niche E, one of the six large niches on the inside wall (Figure 3.2). The beam strikes the wall above the niche at sunrise and slowly moves down to illuminate the niche about 30 minutes after sunrise. The view of the sun from the niche is at such an oblique angle that, although the opening is fairly wide, it projects a somewhat narrow shaft of light into the niche.

The opening will also work for anticipatory observations. Sunlight first breaks through the opening about seven weeks before the summer solstice, an azimuthal change of about 12 degrees from the solstice position. [Note: the errors in all measured angles quoted here are 1-2 degrees, because observations were made by visual sighting with a hand-held compass.] Sunlight initially appears through the opening about May 1. It hits the southwest wall as a narrow shaft that grows broader daily as the solstice approaches. The beam tracks south along the wall as the sun moves northward at an

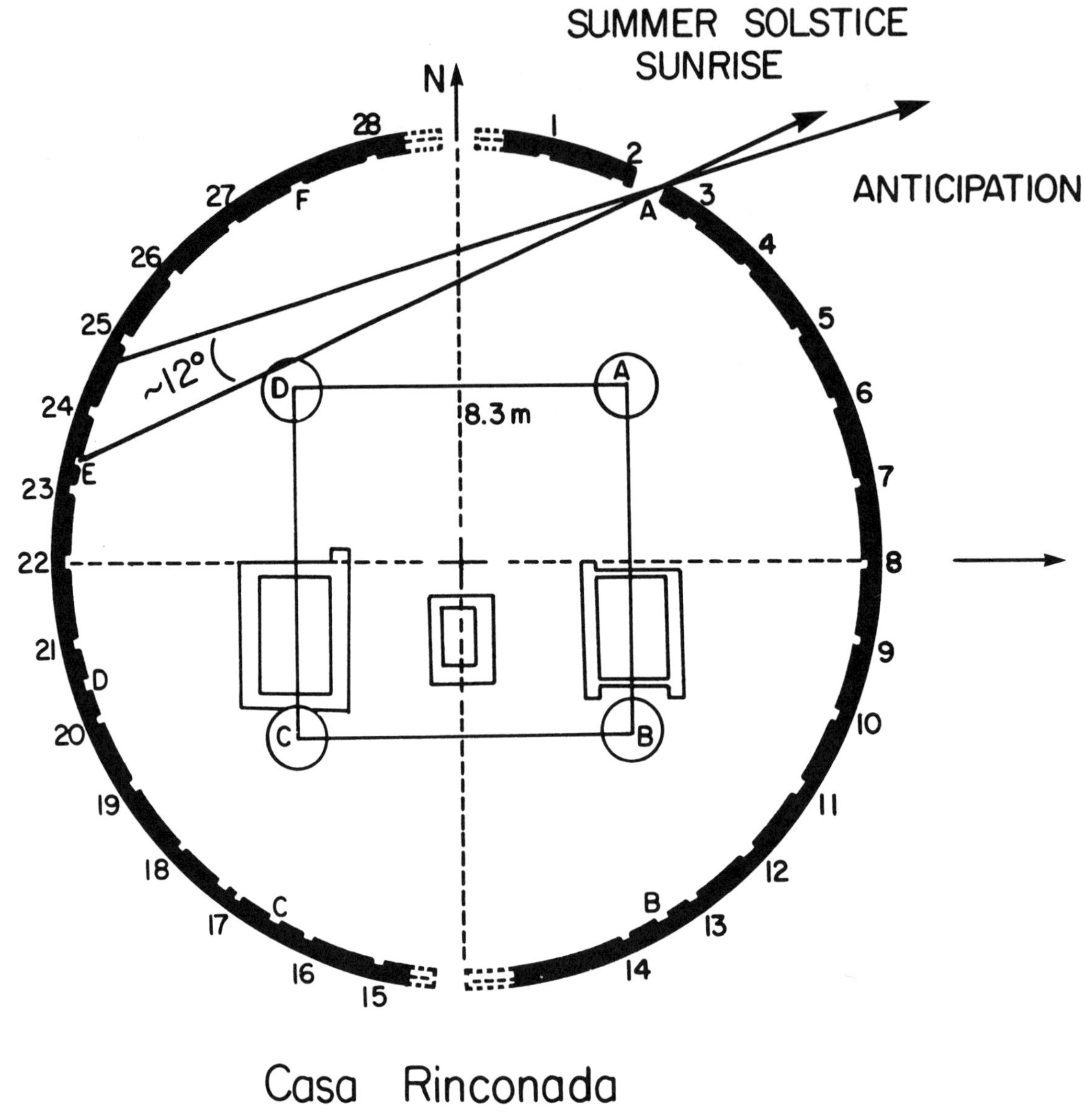

Figure 3.2. The angular relationships of the northwest opening in Casa Rinconada and the niches on the inside of the west side (adapted from Williamson 1982:Fig. 2)

average speed of 6 cm per day. It strikes niche 24 about four weeks before the solstice; this could serve as an anticipatory marker because it could be used to predict the day of the summer solstice.

There are four major problems with the suggestion that Casa Rinconada served as a solar calendar station. First, we do not know the original size and placement of the opening through which the sun appeared prior to restoration by the National Park Service in the 1930s. We do not even know for certain whether a wall existed outside of the window that would have blocked out the sun. An outside room does exist in a similar location at the great kiva of Chetro Ketl (Vivian and Reiter 1965:27-44).

Second, visual examination shows that if roof post D was placed where now indicated and was at least 16 inches in diameter, the post would have blocked the light path to niche E around the summer solstice. One Chetro Ketl roof post found during the excavation was 20 inches in diameter.

Third, I have found only one ethnographic reference to calendrical sun watching from a kiva. Hewett (1938:91-92) notes that the Snake Kiva at Bandelier has a series of unusual postholes in the floor. He was told by Tewa informants that one of the posts was used to track the sun, presumably by the shadow it cast, but these holes were probably used to secure weaving looms.

Fourth, the niches may have been covered over, as were those in Chetro Ketl's great kiva (Vivian and Reiter 1965:29-36), and so they may not have been available for use as calendrical markers, even if all the other problems are resolved. Hence, I suspect that the summer solstice light display is an effect resulting from the partial reconstruction rather than from intention by the Anasazi.

Pueblo Bonito

Reyman (1976) has suggested that the corner doorways in Rooms 225B and 228B at Pueblo Bonito could be used to confirm the winter solstice sunrise. Because both windows are fairly wide (enabling views of a few degrees from the far sides of the rooms), I find it not at all remarkable, given their general southeastern orientation, that the winter solstice sunrise can be seen in the center of the openings from some spot within each room.

I have measured Room 228B and found that the opening could be used for anticipatory observations. The rising sun will first appear in the window at a cross-jamb sighting to the north, 13 degrees away from the winter solstice azimuth (about seven weeks before the solstice). The light will first appear on the wall as a thin beam that will gradually increase in width as the sun moves toward its winter solstice rising point at an average speed of 3 cm/day. Hence, with wall markers, the beam's horizontal motion could be used to anticipate the winter solstice to within a day.

The main problem here, as Reyman (1976) has noted, is that if the outer wall was originally two stories high, it would have blocked out the view of the sun. Reyman (personal communication 1983) states that the old photos and drawings do not show any indication of a second-story wall exterior to these openings, but Pueblo Bonito suffered considerable disturbance in the late nineteenth century and even in prehistoric times after its Chacoan builders had deserted it. So the height of the original outer wall is not yet known, nor is it known whether the corner openings were, in fact, doorways rather than windows.

Fajada Butte

Sofaer and others (Sofaer et al. 1979, 1982) have proposed that three slabs near the top of Fajada Butte cast two shafts of light that can be used to determine accurately the recurrence of the solstices and equinoxes (see Chapter 4). They develop an elaborate case for the intentional positioning and working of the edges of the slabs and also for the positioning of two petroglyphs on the rock face behind the slabs that may act as seasonal markers.

The site clearly shows an interesting play of light between the solstices and the equinoxes, so it has the potential for anticipatory observations. The issue, then, is the practical precision attainable from the site, assuming that it is an intentional calendrical construct used by the Anasazi. The large spiral is about 40 cm in diameter. One month prior to the summer solstice, one of the light shafts falls about 3 cm to the right of the spiral's center; the average daily motion of this light image during the month prior to the summer solstice is about 1 mm/day--the thickness of a dime. The shaft moves horizontally a total of only 2 mm in the four days centered on the summer solstice. In practical terms, I think that these linear motions are too small to be used to anticipate or confirm the summer solstices to within a day or two. As I noted above, Zuni and Hopi sunwatchers were usually able to predict the solstice to within a day of its actual occurrence using horizon calendars. Because of the relatively short horizontal track of the light shafts from winter to summer solstice, the three-slab site at Fajada Butte cannot realistically work at this level of accuracy.

Finally, I will note that it is not common practice in the historical pueblos for the sunwatcher to travel very far to make his observations. They are usually made from within or at the edge of the pueblo. The only significant exception occurs at Zuni, when the pekwin traveled to the Matsakya tower a few kilometers from the present village. That may reflect the historical fact that Matsakya was one of the Zuni villages deserted during the move to the present pueblo location. So the sun-watching station there was established at a time when it was close to the village.

In contrast to the ease of access of present and historically recorded sun-watching stations, the Fajada Butte site would have required a longish walk from the nearest Bonito phase pueblo (Una Vida) and a difficult climb (especially in the winter, when ice and snow coat the chimney). If this site was used by the Chacoan Anasazi, I suspect that it was a sun shrine (such as are historically known on the tops of buttes and mesas) rather than a calendrical sun-watching site.

Hovenweep

Built during the late 1100s and early 1200s, towerlike structures located on rocky outcrops make up some of the ruins at Hovenweep National Monument. Most of these buildings have small ports built into them. These ports are typically 10 cm in diameter and made with considerable attention to their direction of view; a large fraction are angled into the walls. Some still have finished, plastered interiors--probably all did originally. This leads to a cautionary note: Williamson (personal communication, 1983) has emphasized that any interior plastering would cut down the angular view and so shorten the times and tracking distances mentioned below.

In contrast to Chaco, the Hovenweep ruins have not been reconstructed, though they have been stabilized. An examination of the photos gathered by Fewkes (1919:Plates 14a, b, and c, 19c, and 32b) shows that in the late nineteenth century the ruins appeared very much as they do today, except that some of the walls have fallen since then.

In three of the ruins--Hovenweep Castle, Unit Type House, and Cajon Ruin--rectangular rooms were built as late additions to each structure. That these rooms were later additions is inferred from their different masonry style and the lack of integration between these rooms and the rest of the structure. Williamson and others (Williamson 1981; Williamson et al. 1977) have proposed that ports in these add-on rooms were used for solar observations, with alignments for the solstices and equinoxes. From an astronomical view, the ports define arrow angular views on the horizon and permit narrow shafts of light to enter the rooms.

In Hovenweep Castle, the possible solar observation room is situated on the south end of a D-shaped tower. Two ports cut through different walls: one opens to the northwest and provides an entrance for light from the summer solstice sunset; the other opens to the southwest toward the winter solstice sunset (Figure 3.3). Light from the summer solstice sunset falls on the north side of the lintel of the east door. About 70 days before the summer solstice, the sunlight first enters the room with the beam striking the west side of the lintel of the north door. As the sun moves north, the sunset point tracks from one lintel to the other, a linear distance that corresponds to an angle of roughly 17 degrees on the horizon (Figure 3.3).

A similar pattern occurs for the light entering the winter solstice port. At the winter solstice sunset, the light strikes the east side of the north doorway's lintel. The beam first appears 70 cm to the right of this position, or an azimuthal angle of about 17 degrees. So the sunset beam enters the room some 70 days before the winter solstice. If markers were placed on the inside north wall (a practice well known for Zuni [Cushing 1979:117]), the tracks of the sunlight would make it quite easy to anticipate either solstice with a day tally or count.

Markers could also have been used to track a planting calendar using the summer solstice port. The climate at Hovenweep, and so the planting/harvesting cycle, is very similar to that of the Hopi mesas. At Hopi, seeds are first planted around April 15, and planting continues until the summer solstice (Beaglehole 1937; Forde 1934). At Hovenweep Castle the sunlight first enters the summer solstice port around April 7--just in time to set the start of the planting season. So this port could have served three purposes: to anticipate and confirm the summer solstice and to establish the schedule for crop planting. Note that the oblique angle to the wall creates a long track for the planting season interval.

At the Cajon Group ruins, about 10 km south of Hovenweep Castle, the possible solar observation room contains three ports, each roughly 10 cm across. They face west (and thus function only at sunset). The two end ones are perhaps winter and summer solstice observational ports. The center one provides an observation point for the equinox sunsets (Figure 3.4). During the time when

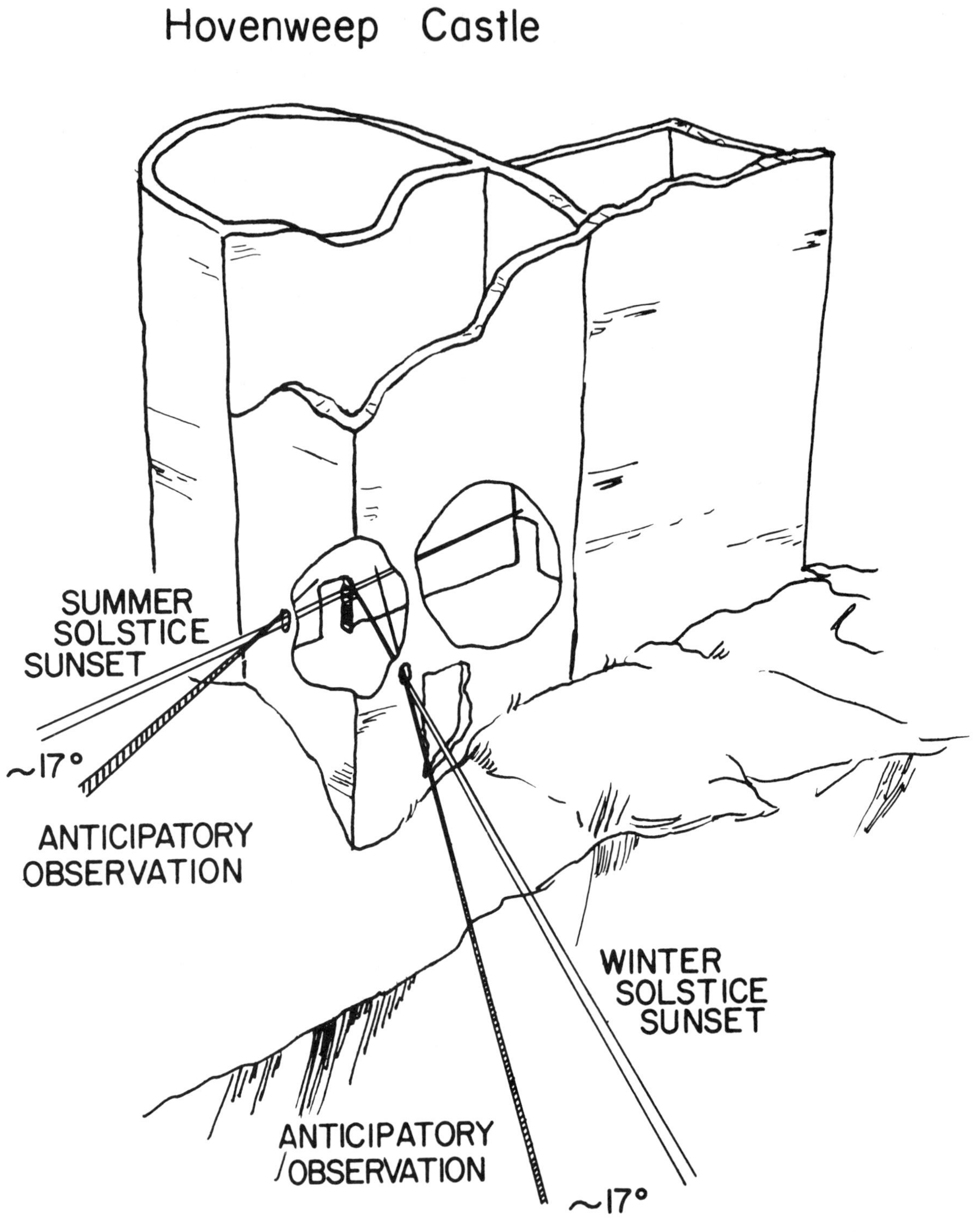

Figure 3.3. The orientations of the sunset light entering the winter and summer solstice ports at Hovenweep Castle (adapted from Williamson 1981:Fig. 8.5)

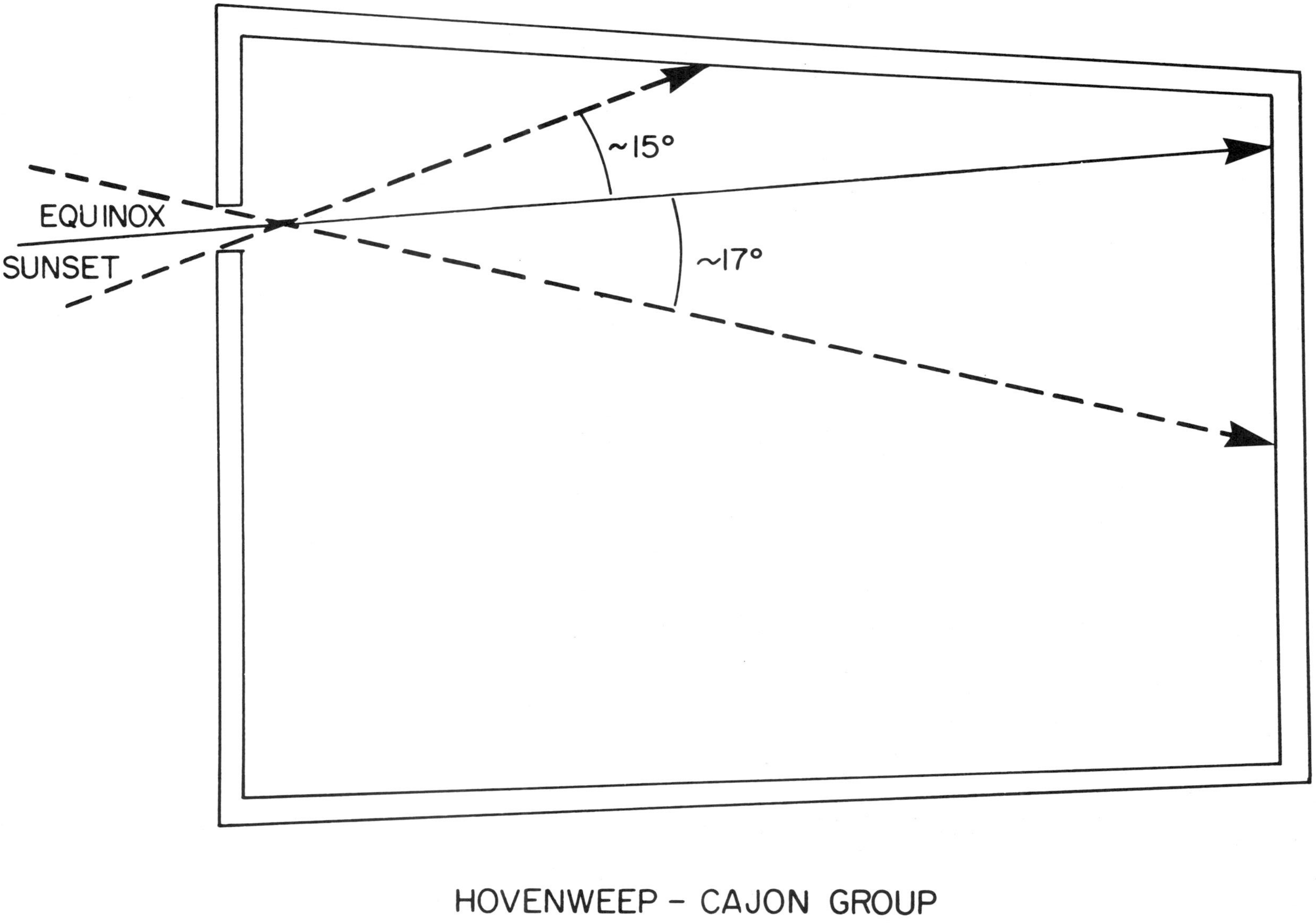

Figure 3.4. The orientations of sunlight entering the equinox portal of the possible observing room at Cajon Group, Hovenweep

the sun moves north, light first enters this port 32 days before the spring equinox and disappears 36 days afterward. The reverse occurs when the sun moves south around the time of the fall equinox. The average linear motion of the beam is roughly 3 cm/day. Because the historical Pueblos do not celebrate the equinoxes as such, we need to consider planting dates as well. At Hopi, the planting season begins approximately one month after the spring equinox (Forde 1931). In one hypothetical scenario, the middle port could be viewed as providing the signal for the start of the planting season. The northern port then may provide the timing for the sequence of plantings, as well as anticipating the summer solstice. Additional fieldwork is needed in order to support or refute this hypothesis.

The possible solar observation room at Unit Type House contains four ports, one of which Williamson et al. (1977) have suggested sights onto the southernmost declination of the moon at major standstill. This room has suffered more deterioration than the observatory rooms at Hovenweep Castle and Cajon. The thick layer of debris on the floor interferes with observations of the sunlight beams, so the phenomena described are less certain than those described for the other Hovenweep sites.

Light from the summer solstice sunrise enters a port and shines on the southwest corner of the room (Figure 3.5) about one-half hour after sunrise. If junipers along the horizon do not interfere too much, the sunlight first appears through this port about 12 weeks before the solstice and hits the northwest room corner. The light then tracks completely across the wall. Similarly, at solstice the winter solstice portal admits sunlight into the northwest corner, while anticipatory observations could begin nine weeks before, when the beam first enters, hitting the center of the west wall.

ANTICIPATION AND ACCURACY

At Hopi and Zuni, the pueblos least influenced by the Spanish, the sun priests can usually predict the solstices to the day using horizon markers. (The situation is more complicated at Zuni because of the desire to conjoin the winter solstice with a full moon and the summer solstice with a new moon [Tedlock 1983]). Let us assume, then, an error of plus or minus one day as the standard of accuracy to apply to prehistoric sites.

For those possible calendrical sites that depend on a beam of light to mark the desired astronomical event, the linear distance that the beam moves per day determines the level of accuracy that can be obtained. I suggest that 1 cm/day is the minimum speed for one-day precision in prediction. Table 3.1 gives the total track lengths, duration of anticipation, and average motion of the beam per day. [Note: Keep in mind that as the solstices approach, the motion of the beam slows considerably.] This analysis shows that the average rates range from about 1 to 5 cm/day, most of which are equal to or greater than the minimum detectable motion to permit a one-day accuracy of prediction. The sun shaft on the right at the Fajada Butte site, because its total horizontal track is so short, fails to meet this standard by at least a factor of 10. At best this site could have served as an approximate confirmatory solstice marker, if in fact it was used as a calendrical site by the Anasazi.

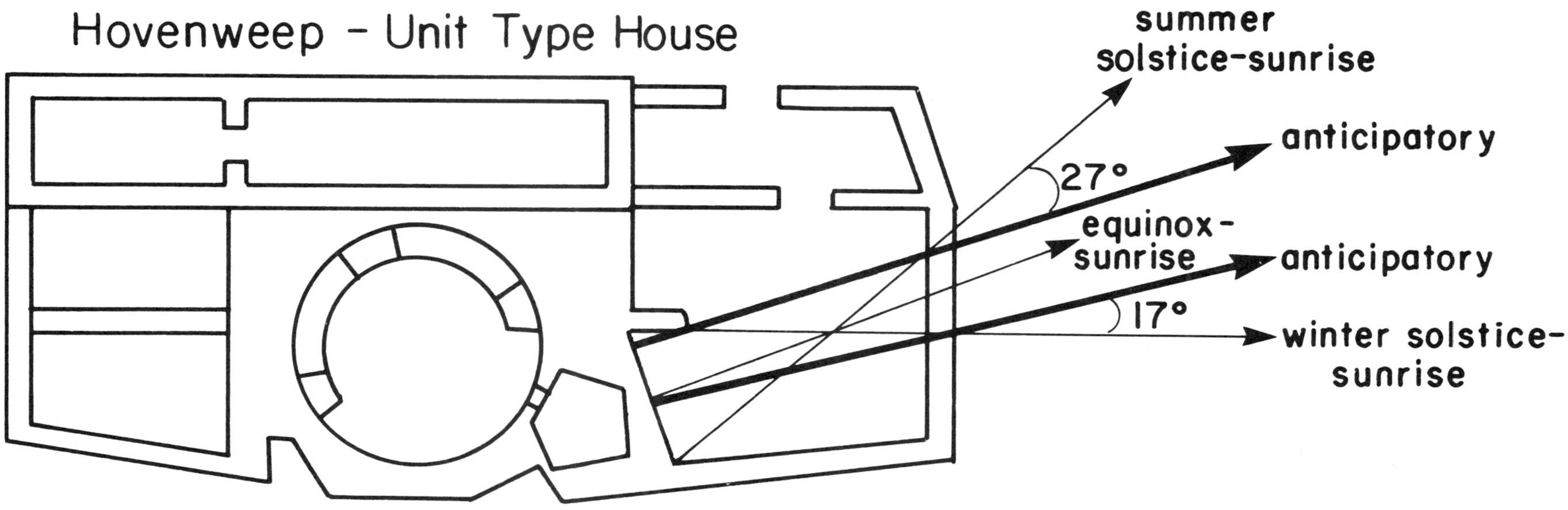

Figure 3.5. Angular orientations of the sunrise light entering the solstice and equinox portals at Unit Type House, Hovenweep

Table 3.1. Anticipation of solar events at Anasazi sites

Site	Total Track (cm)	Duration (days)	Average Speed (cm/day)
Hovenweep Castle			
Summer solstice port	224	67	3.6
Winter solstice port	56	70	0.9
Unit House			
Summer solstice port	211	87	2.4
Winter solstice port	201	57	1.8
Cajon Ruin			
Equinox port	216	68	3.2
Pueblo Bonito			
Room 228	165	53	3.1
Casa Rinconada			
Northeast "window"	300	50	6.0
Fajada Butte (right-hand light shaft)			
Summer solstice to equinox	7	90	0.08
Summer solstice to winter solstice	18	180	0.1

OVERALL EVALUATION OF ANASAZI SOLAR SITES

I raise again the issue of the intentionality of prehistoric astronomical sites and suggest that any purported astronomical sites need to be evaluated in a cultural as well as an astronomical context; it is not enough for the site to have alignments with equinox and/or solstice solar phenomena. I contend that calendrical sites should be useful for anticipatory observations and perhaps as aids in planning a planting sequence--both key functions performed by the historical sun priests. This argument does not apply to sun shrines, however, which have a ceremonial function as a place to present offerings.

I have made a preliminary attempt at an overall, relative evaluation of the sites discussed here based on five criteria: site integrity (reconstruction, disturbance, etc.), archaeological context, confirmatory use, anticipatory use, and ethnographic context. The ratings in Table 3.2 are somewhat subjective and only valid in a relative sense. I have assigned a "+" where the evidence to date supports the usefulness of the site; a "0" where the data are contradictory, disputed, or incomplete; and a "-" for evidence that swings to the negative side. This simple analysis shows that the Hovenweep sites have, overall, the strongest probability of being intentional sun-watching sites.

ACKNOWLEDGMENTS

I could not have covered the space or completed the fieldwork for this project in the time available without the help of Sara Parker, Alberto Enriquez, Bryan Laubscher, and Janet Saad-Cook. I also thank Jim Judge, Jane Young, Ray Williamson, Joan Mathien, Jonathan Reyman, and Bob Hart for constructive discussions. Special thanks go to Randy Morrison at Chaco Canyon for his hospitality and intelligent help. Finally, Florence Ellis deserves my gratitude for her careful review of the manuscript.

Table 3.2. Evaluation of possible Anasazi sun-watching sites

Site	Integrity	Archaeological Context	Confirmatory	Anticipatory	Ethnographic Context
Hovenweep	+	+	+	+	+
Pueblo Bonito	0	+	+	+	+
Casa Rinconada	0	+	+	+	0
Fajada Butte	0	0	+	0	0

REFERENCES CITED

Beaglehole, E.
1937 Notes on Hopi Economic Life. Yale University Publications in Anthropology 15:1-88.

Benedict, R. F.
1935 Zuni Mythology. Columbia University Contributions to Anthropology 21(II):62-68.

Cushing, F. C.
1979 My Adventures in Zuni. In Zuni: Selected Writings of Frank Hamilton Cushing, edited by J. Green, pp. 46-134. University of Nebraska Press, Lincoln.

Dutton, B. P., and M. A. Marmon
1936 The Laguna Calendar. University of New Mexico Bulletin, Anthropological Series 1(2):1-21.

Ellis, F. H.
1975 A Thousand Years of the Pueblo Sun-Moon-Star Calendar. In Archaeoastronomy in Pre-Columbian America, edited by A. F. Aveni, pp. 59-87. University of Texas Press, Austin.

Fewkes, J. W.
1919 Prehistoric Villages, Castles, and Towers of Southwestern Colorado. Bureau of American Ethnology Bulletin 70:1-79.

Forde, C. D.
1931 Hopi Agriculture and Land Ownership. Journal of the Royal Anthropological Institute of Great Britain and Ireland 51:357-405.
1934 Habitat, Economy and Society. E. P. Dutton, New York.

Hewett, E. L.
1938 Pajarito Plateau and Its Ancient People. University of New Mexico Press, Albuquerque.

Meeus, J.
1983 Astronomical Tables of the Sun, Moon, and Planets. Willmann-Bell, Richmond, Virginia.

1925 A Pueblo Indian Journal, 1920-1921. Memoirs of the American Anthropological Association 32.

Parsons, E. C.
1932 Isleta, New Mexico. Bureau of American Ethnology Annual Report 47:193-466.

Parsons, E. C., editor
1936 Hopi Journal of A. M. Stephen, 2 vols. Columbia University Press, New York.

Reyman, J. E.
1971 Mexican Influence on Southwestern Ceremonialism. Ph.D. Dissertation, Department of Anthropology, Southern Illinois University, Carbondale.
1976 Astronomy, Architecture, and Adaptation at Pueblo Bonito. Science 193:957-962.
1980 The Predictive Dimension of Priestly Power. In New Frontiers in the Archaeology and Ethnohistory of the Greater Southwest, edited by C. L. Riley and B. C. Hendrick, pp. 40-59. Transactions of the Illinois Academy of Science 72(4).

Sofaer, A., R. M. Sinclair, and L. E. Doggett
1982 Lunar Markings on Fajada Butte, Chaco Canyon, New Mexico. In Archaeoastronomy in the New World, edited by A. F. Aveni, pp. 169-181. Cambridge University Press, Cambridge.

Sofaer, A., V. Zinser, and R. M. Sinclair
1979 A Unique Solar Marking Construct. Science 206:283-291.

Tedlock, B.
1983 Zuni Sacred Theater. American Indian Quarterly 7(4):93-110.

Vivian, G., and P. Reiter
1965 The Great Kivas of Chaco Canyon and Their Relationships. School of American Research Monograph 2. Santa Fe.

Williamson, R. A.
1981 North America: A Multiplicity of Astronomies. In Archaeoastronomy in the Americas, edited by R. A. Williamson, pp. 61-80. Ballena Press, Los Altos, California.
1982 Casa Rinconada, Twelfth Century Anasazi Kiva. In Archaeoastronomy in the New World, edited by A. F. Aveni, pp. 205-219. Cambridge University Press, Cambridge.

Williamson, R. A., H. J. Fisher, and D. O'Flynn
1977 Anasazi Solar Observatories. In Native American Astronomy, edited by A. Aveni, pp. 203-217. University of Texas Press, Austin.

Williamson, R. A., H. J. Fisher, A. F. Williamson, and C. Cochran
1975 The Astronomical Record in Chaco Canyon, New Mexico. In Archaeoastronomy in Pre-Columbian America, edited by A. F. Aveni, pp. 33-43. University of Texas Press, Austin.

Zeilik, M.
1983a Anticipation in Anasazi Astronomy. Paper presented at the 56th Annual Pecos Conference, August 1983, Bluff, Utah. Ms. in author's possession.
1983b Historic Pueblo Sun Watching. Paper presented at the First International Ethnoastronomy Conference, September 1983, Smithsonian Institution, Washington, D. C.

Chapter 4

Astronomical Markings at Three Sites on Fajada Butte

Anna P. Sofaer and Rolf M. Sinclair

A seven-year study of prehistoric Pueblo sites of the Chaco culture has revealed that these people possessed a sophisticated astronomy. Evidence of this prehistoric astronomy consists of multiple light markings on petroglyphs and of several alignments of major structures. This paper presents the results of a recent study of 13 markings at three sites on Fajada Butte in Chaco Canyon, New Mexico; these results are presented in the context of earlier research in this area. Each marking is a distinctive pattern of shadow and light that appears on a petroglyph at a key point in the solar or lunar cycle, i.e., at an extreme or midposition of these cycles, including the meridian passage of the sun at solar noon. [Note: times quoted throughout this chapter are in apparent solar time, in which noon occurs each day when the sun is due south on the meridian. The basic astronomical concepts used here are explained in Aveni (1980) and Krupp (1977).] Many of the markings simultaneously record two key points in different cycles, such as noon and solstice or noon and equinox. A minimum of 17 key points are indicated by the markings at these three sites.

The markings define the outer boundaries and midpoints of the recurring cycles of the two most basic celestial bodies: the sun and the moon. The accuracy and redundancy of the markings of these symmetric points, as well as the strong visual effect of the markings themselves, indicate their intentional quality. Several of the markings record meridian passage of the sun within a few minutes, and one marks equinox to within a day. An accuracy equivalent to that of these markings is found in the cardinal alignments of major constructions in the area.

Fajada Butte stands 135 m high at the south entrance of Chaco Canyon. The remains of various structures, including a small kiva, and the presence of potsherds and rock art on and near the butte summit show that it was frequented by prehistoric Pueblo people and by Navajo people. The content and style of the glyphs at the three sites reported here indicate that these markings are of prehistoric Pueblo origin and were probably made between AD 900 and 1300.

From about AD 900 to 1150 Chaco Canyon was the center of a complex prehistoric Pueblo society that thrived in the arid environment of northwestern New Mexico. The Chacoan people planned and constructed large multistory structures, ceremonial centers, and extensive roads throughout the 70,000 sq km San Juan Basin. These achievements indicate sophisticated engineering and surveying skills. Ethnographic reports concerning the historical Pueblos (some of whom are descendents of the Chacoan people) reveal their keen interest and skills in observing the sun and moon for ritual and agrarian purposes (see references cited in Sofaer et al. 1979).

The Fajada markings, although similar to others recently reported in the Southwest, are unique among known archaeoastronomical sites in several respects: (a) they use the changing altitude of late morning and midday sun to indicate the solstices and equinoxes, (b) they record solar noon on rock art, and (c) most especially, they combine recording of both the season and noon in the same markings. The markings at one site are the only known ones in the New World that combine recordings of the extremes and midpositions of the moon and the sun. The use of the features and topography of a prominent land mass to create these complex and varied markings on carved petroglyphs is also unique in our current knowledge of prehistoric astronomy.

Following a brief summary of earlier findings at one of the three sites near the top of Fajada Butte (see also Sofaer et al. 1979, 1982a), this chapter describes in detail the recently discovered markings that combine noon and seasonal recordings at two additional sites on the east and west sides of the butte (Sofaer et al. 1982b). New observations and analyses regarding the earlier findings are then presented. The sites are referred to in this paper as the three-slab site, the east site, and the west site. The markings of these sites were first observed and studied by Sofaer between 1978 and 1983. The markings are summarized in Table 4.1. Ethnographic and archaeological correspondences with the markings are also discussed.

SUMMARY OF EARLIER FINDINGS

At the three-slab site, six markings record solar and lunar positions (Figures 4.1, 4.2, 4.3). An unusual configuration of three large stone slabs, each about 2 m high, collimates sunlight each day in the late morning and near midday onto two spiral petroglyphs pecked on a cliff face. The streaks of light so formed change noticeably with small changes in the sun's declination. The vertical light/shadow patterns on the petroglyphs thus go through an annual cycle in which the solstices and equinoxes are marked by the intersection of these patterns with the primary features of the spiral forms (Sofaer et al. 1979). This site also marks the northern minor and major extremes of the 18.6-year lunar standstill cycle by a separate pattern of light and shadow at moonrise (Sofaer et al. 1982a). This light and shadow pattern marks events that are unrelated to those of the solar cycle, yet it is also formed on the primary features of the larger spiral by another edge of one of the three slabs that form the midday patterns.

Since the moon can appear throughout the declination range of the sun, all the solar markings listed in Table 4.1 can also be formed at certain times by the moon. We do not consider most of these to be possible lunar markings owing to a lack of further evidence. The northern minor standstill marking can be formed at certain sunrises away from the solstices and equinoxes; we term it lunar because it occurs as a pair with the northern major standstill marking (which cannot be solar since the moon then

Table 4.1. Markings of astronomical cycles on Fajada Butte

	SOLAR					LUNAR	
	Midday				Sunrise	Moonrise	
Petroglyph	Noon	Summer Solstice	Equinox	Winter Solstice	Equinox	Northern Major Standstill	Northern Minor Standstill
Large spiral (three-slab site)	-	X	-	X	X	X	X
Small spiral (three-slab site)	-	*	X	-	-	-	-
East spiral	X	X	X	X	-	-	-
East snake	X	-	X	-	-	-	-
East rectangle	X	X	-	-	-	-	-
West double spiral	X	-	X	-	-	-	-
West rectangle	X	-	-	-	-	-	-

* The <u>absence</u> of light at this location can be taken as a marking (see text)

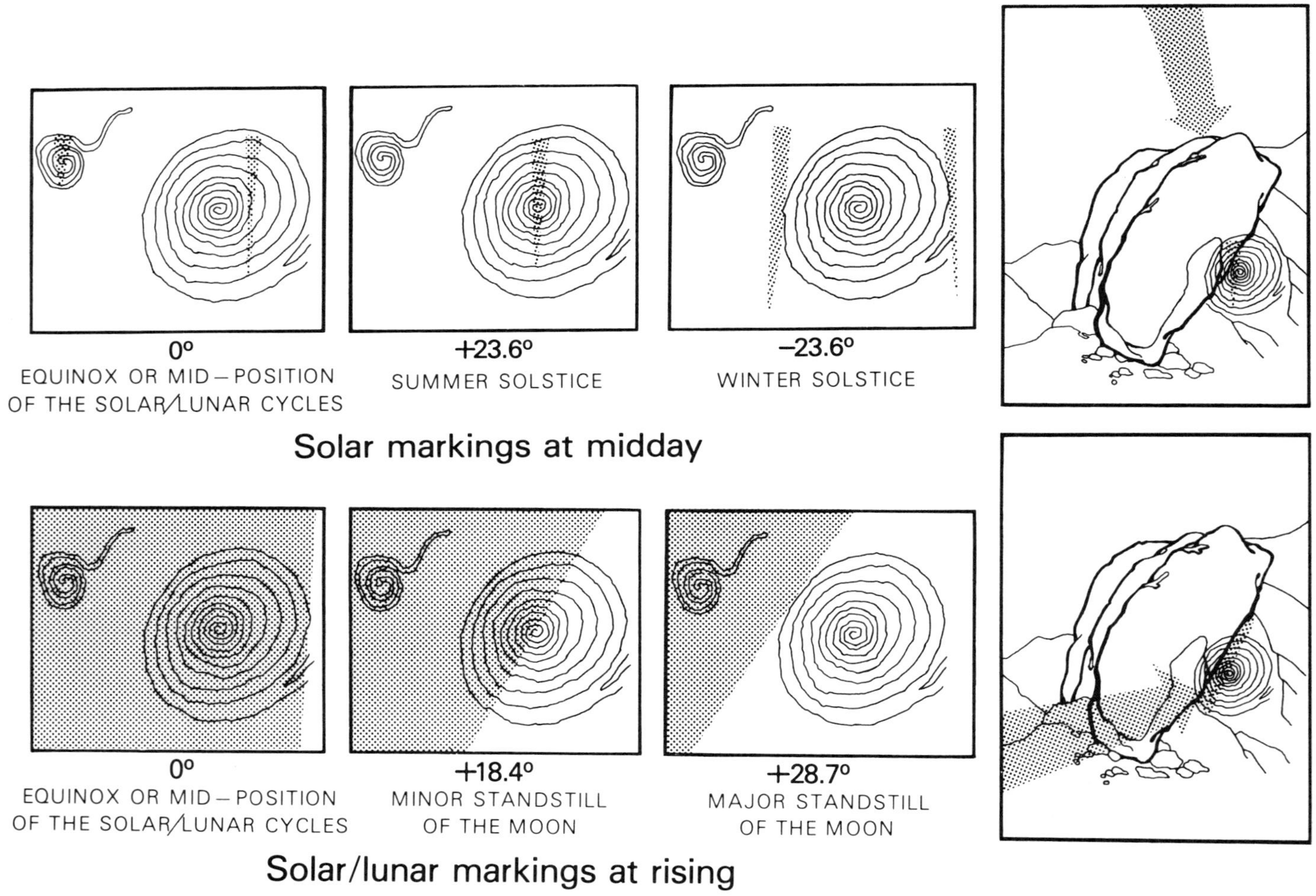

Figure 4.1. The three-slab site. Right: Formation of solar and lunar shadow/light patterns by the three slabs near meridian passage (upper) and at rising (lower). Left: Schematic of the resulting patterns on the spirals at the indicated declinations and seasons (see Sofaer et al. 1979, 1982a for detailed photographs). (Note: an error in an earlier publication [Sofaer et al. 1982a:Fig. 4] of "rising 0°" is corrected here.) Illustration by Pat Kenny, © The Solstice Project

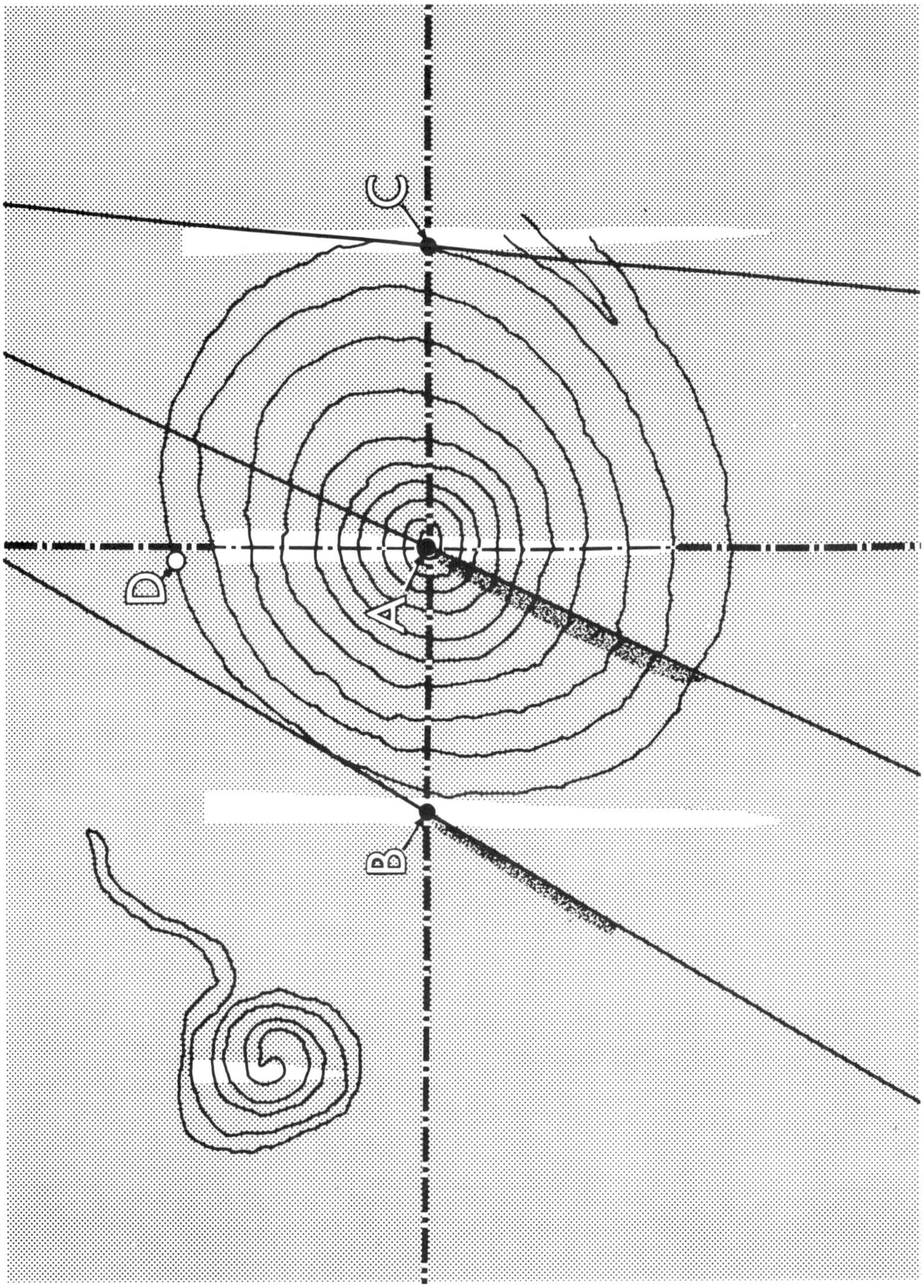

Figure 4.2. The three-slab site. Superposition of the main elements of the six markings (see Figure 4.1) showing the multiple uses of certain key features of the large spiral (A), points of tangency at left and right (B and C), the top (D), and the horizontal and vertical axes. Also shown are the two pecked grooves (see text). Illustration by Pat Kenny, © The Solstice Project

exceeds the maximum solar declination). The equinox sunrise marking (Figure 4.1) also indicates the middle of the moon's monthly or 18.6-year declination cycles, in a manner analogous to the markings of the standstill extremes, and is thus possibly also a lunar marking.

While the solar markings of the three-slab site use the seasonal altitude changes of the late-morning and midday sun to indicate the specific time of the solstices and equinoxes, they do not accurately mark noon. New evidence reveals that two other sites on the butte mark the specific time of noon at the seasonal points. Several features of the three-slab site are shared by the two other sites: spirals, dagger-shaped light patterns, and the repeated use of the same glyph or pair of glyphs at quarter points of the seasons.

RECENT FINDINGS OF NOON/SEASONAL MARKINGS

At two sites located a short distance below the three-slab site (Figure 4.4), five petroglyphs are crossed by visually compelling patterns of shadow and light at a time close to solar noon. The imagery of these patterns distinguishes the solstices and equinoxes in most instances. These shadow patterns form seven markings indicating 11 key points in the daily and seasonal cycles of the sun--i.e., midpositions and extremes (Table 4.1). The markings occur in combinations, most of them in striking conjunctions: pairs that are visible at the same site and that vary with the seasons.

The shadows that form the noon markings are cast by rock edges of the butte, which is locally very irregular. Since the sun's elevation at meridian passage changes annually from 31 degrees at winter solstice to 78 degrees at summer solstice, the particular edges that cast shadows on the glyphs through the year differ greatly in distance and bearing from them. Some edges are less than a meter away, while others are up to 30 m distant. During the sun's daily meridian passage, the glyphs at the east site change from fully lit in morning sun to fully shadowed in the afternoon, and those at the west site go through the reverse transition of shadow to light.

At the east site (Figures 4.5 and 4.6), which is located about 25 m below the butte summit, three adjacent glyphs--a nearly vertical rattlesnake (22 cm long), a rectangular figure (14 cm wide), and a spiral (15 cm wide)--occur 2.0-2.5 m above the current cliff base. The rattlesnake and the rectangular figure are particularly deeply incised. The spiral is pecked within a rectangular area that appears to have been worked. A shadow edge crosses the spiral glyph within 10 minutes of noon throughout the year, forming a seasonally changing pattern. This pattern is momentarily symmetric about the center of the spiral within a few minutes of noon, forming a wedge at summer solstice, a quartering at equinox, and a bisecting at winter solstice.

The seasonal variation is reinforced by simultaneous markings that occur on the two nearby glyphs only within a few weeks of summer solstice and equinox, close to noon. At equinox, at the same time that the spiral is quartered, an unrelated shadow edge crosses the snake, touching all parts at once: the head, the body, and the rattles. This shadow pattern deviates noticeably from alignment with the full body of the snake two weeks earlier or later (Figure 4.6). At summer solstice, at the time that the wedge marks the center of the spiral, the shadow edge is aligned with the

Figure 4.3. Summer solstice light marking on the larger spiral of the three-slab site. The arrows indicate the two pecked grooves with which the lunar shadow patterns are aligned. Photograph by David Brill, © 1982, The National Geographic Society.

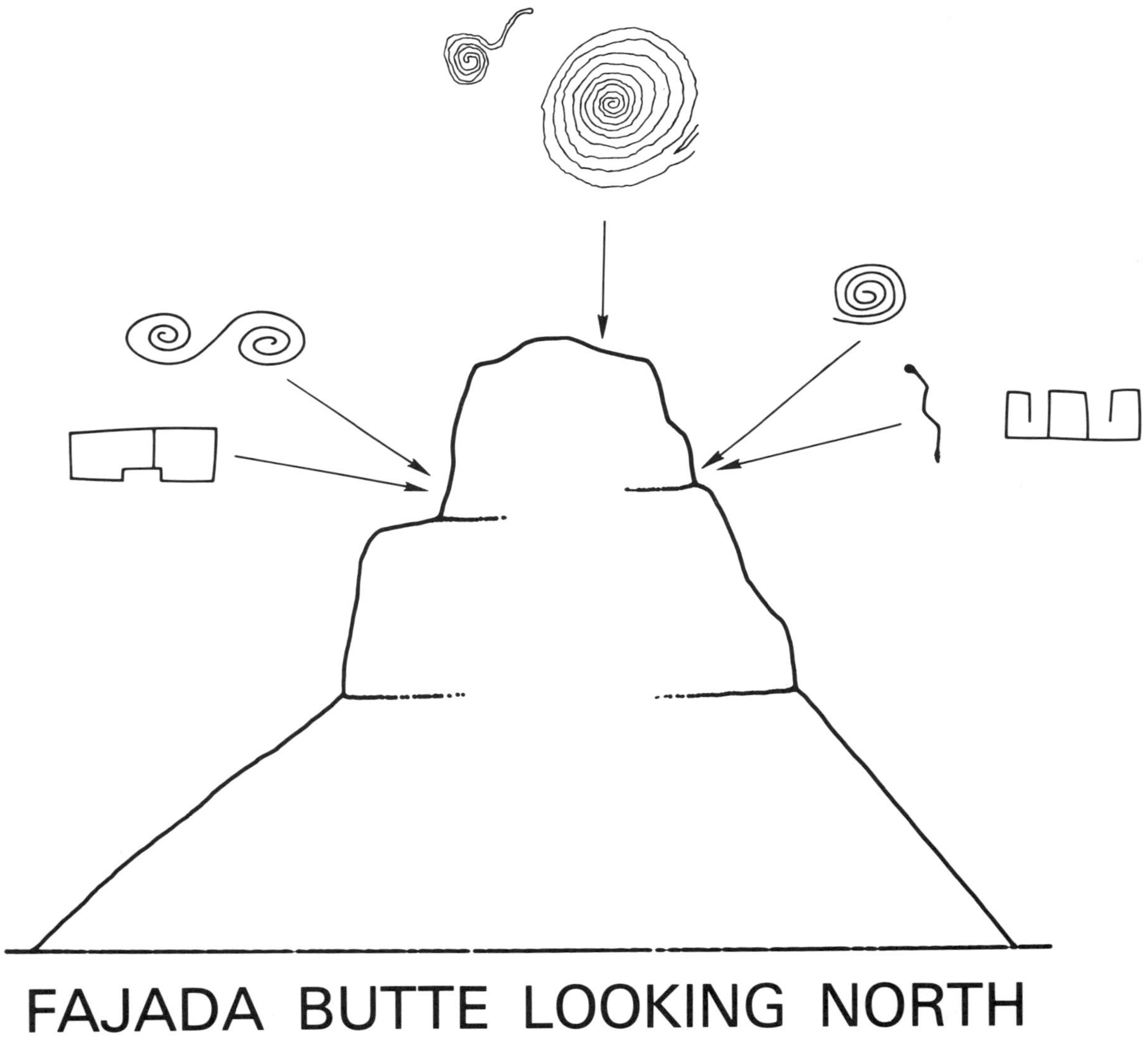

Figure 4.4. Schematic of Fajada Butte (looking north) showing the locations of the petroglyphs at the east, west, and three-slab sites (which are not intervisible). Illustration by Pat Kenny, © The Solstice Project

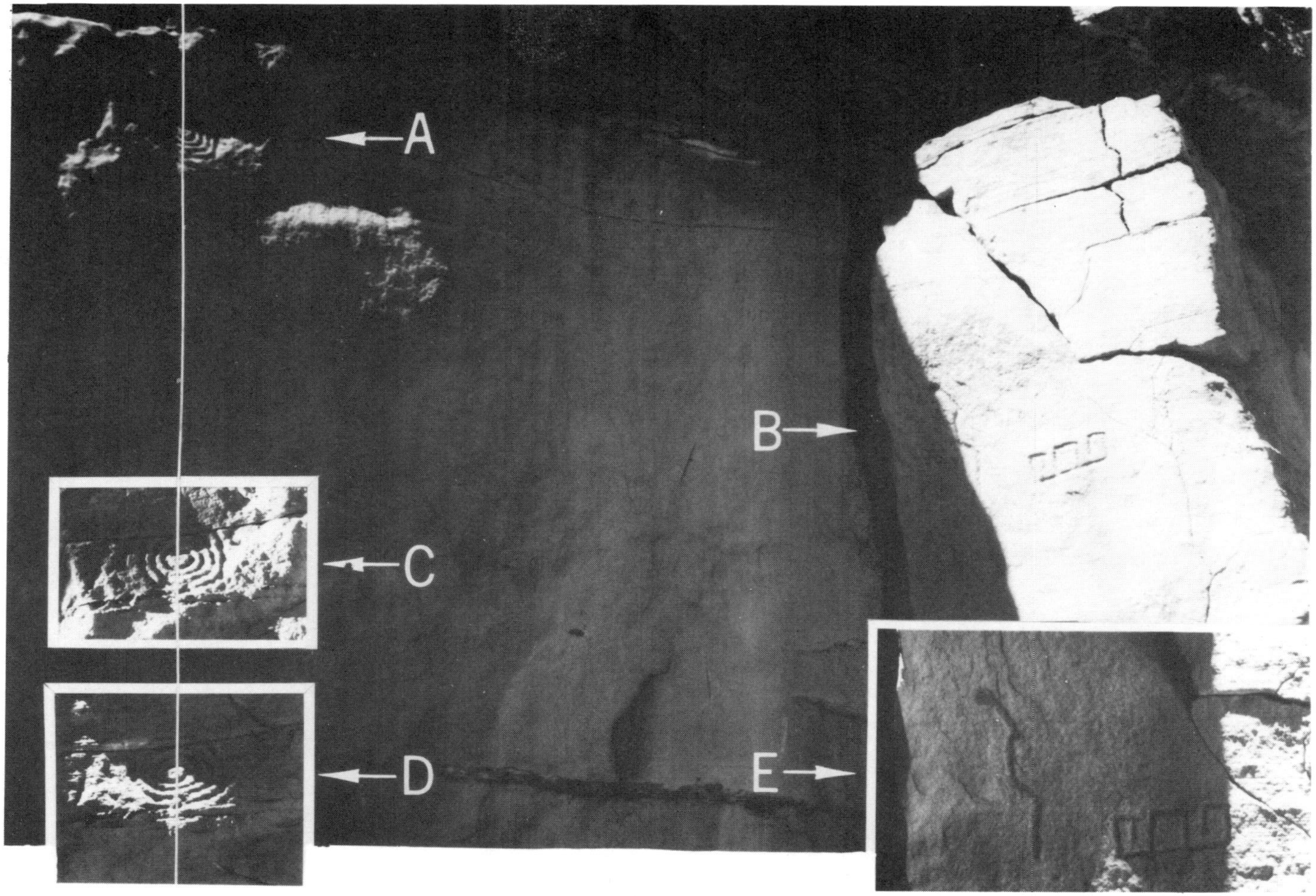

Figure 4.5. East site. Simultaneous markings at equinox on (A) spiral and (B) snake (September 20, 1979, 12:04 pm). Insets: (C) winter solstice marking on spiral (December 20, 1980, 12:00.5 pm), (D and E) summer solstice simultaneous markings on spiral and rectangular figure (June 22, 1978, 12:03 pm). The line through ACD points out the change in the shape of the shadows that centrally mark the spirals at each season around the time of noon. Photograph by A. Sofaer and K. Kernberger, © The Solstice Project

A B

Figure 4.6. East site. Markings on the snake showing the alignment of the shadow edge with the petroglyph at equinox. (A) 16 days after fall equinox (October 9, 1982, 12:00 m), (B) 3 days before fall equinox (September 20, 1979, 2:03.75 pm). [Note: data taken a certain time after (or before) spring equinox are equivalent to data taken the same time before (or after) fall equinox.] Photographs by A. Sofaer, © The Solstice Project right edge of the rectangular figure (Figure 4.5). This combination is noticeably different two weeks before and after summer solstice. Thus, the seasonal points of the year are indicated both by the differing patterns on the individual glyphs and, at summer solstice and equinox, by the unique paired combinations of the markings.

At the west site (Figures 4.7-4.9), which is located about 30 m below the summit, a double spiral (55 cm across) and a rectangular figure (17 cm wide) are about 6 m and 4 m, respectively, above the current cliff base. On days near equinox, close to noon, a narrow daggerlike pattern of light moves in an upward diagonal course through the double spiral in about 22 minutes. An unrelated vertical shadow edge moves across the rectangle in nearly the same interval. At equinox the light pattern moves through the center of the right whorl of the double spiral, with the whorl and the rectangle being bisected simultaneously about 9 minutes before noon. The moving light patterns reach the far right edges of both glyphs within a few minutes of noon.

Movement of light across the double spiral near equinox is a sensitive indicator of the sun's declination (which is then changing most rapidly). The track of the pattern shifts 4.5 cm each day, left to right in the spring and right to left in the autumn, so that the pattern of light movement for each day is quite distinct during this period. Figure 4.9 shows the light pattern on several dates near equinox. In the photographs of the double spiral near equinox, the light form is shown at the same height in its upward diagonal course for each day. The "event" of equinox can occur through the years at different times on the nominal "day" of equinox. Figures 4.7 and 4.8 show that a shift in successive years of only 12 hours in the time of equinox can be seen. By noting the exact position of the light pattern with respect to the center of the right whorl, one can pinpoint equinox to within at least a half day.

The shift of the track of this light pattern from left to right on the double spiral could be used to anticipate the approach of spring equinox [Chapter 3, this volume], since even to a casual observer a one- or two-day shift from equinox would be evident. While no external evidence is available to determine whether this site was used to mark the time of equinox so accurately, it was clearly possible.

The markings on the rectangle are of special interest in that they indicate noon accurately throughout the year. Each day a vertical shadow edge crosses the rectangle within 5-20 minutes of noon and aligns momentarily with the right edge within 3 minutes of noon throughout the year.

Figure 4.10 illustrates the intervals around noon between first and last touchings by a shadow edge on each glyph at both the east and west sites. This figure also shows certain times within these intervals when several of the glyphs are marked by distinctive and centrally aligned patterns; these are illustrated in Figures 4.5-4.9 as well. Several of these patterns occur within a few minutes of noon.

Several factors underscore the intentional quality of the markings described here. First, in all instances the light patterns bisect or align with the edges of the glyphs at key points in the solar cycles. Second, six of the seven markings record two such events (noon and a solstice or equinox). Third, the visual effect is striking, most

Figure 4.7. West site, 1.5 hours before fall equinox (September 22, 1984, 11:50.5 am). The shadow edge moves across the rectangular glyph (lower right) from left to right, crossing the right edge within three minutes of noon throughout the year. (The outline of this glyph is artificially emphasized in this illustration.) Photograph by Colin Franklin, © The Solstice Project

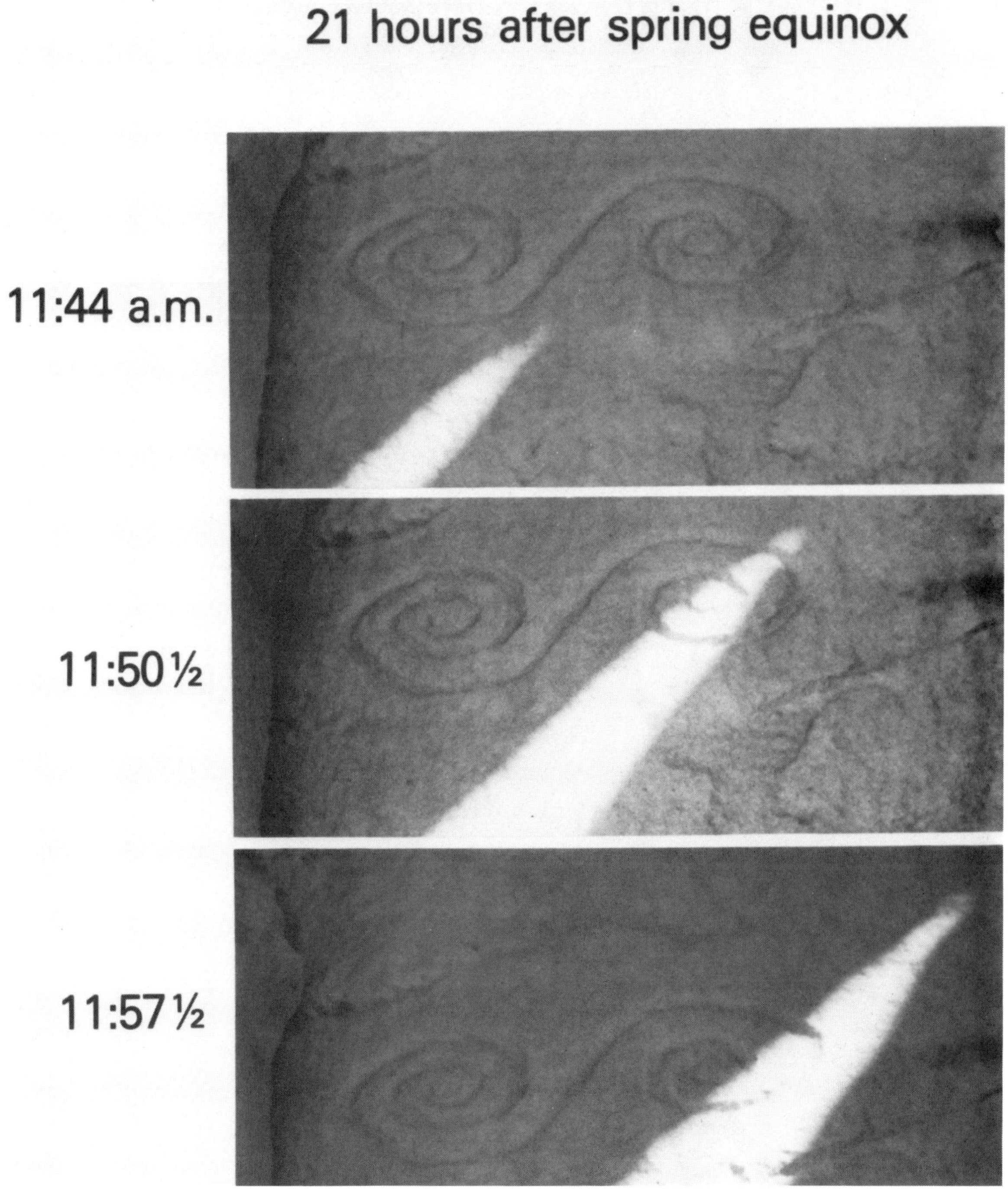

Figure 4.8. West site. Marking on double spiral 21 hours before fall equinox (September 22, 1983, top to bottom: 11:44 am, 11:50.5 am, 11:57.5 am). Photographs by Peggy Wier, © The Solstice Project

Figure 4.9. West site. Marking on double spiral near equinox. Top: September 30, 1984, 11:42.5 am (8 days after fall equinox). Middle: March 20, 1984, 11:48.5 am (9 hours after spring equinox). Bottom: March 23, 1984, 11:53.5 am (3 days after spring equinox). Photographs by Rolf Sinclair and Michael Marshall, © The Solstice Project

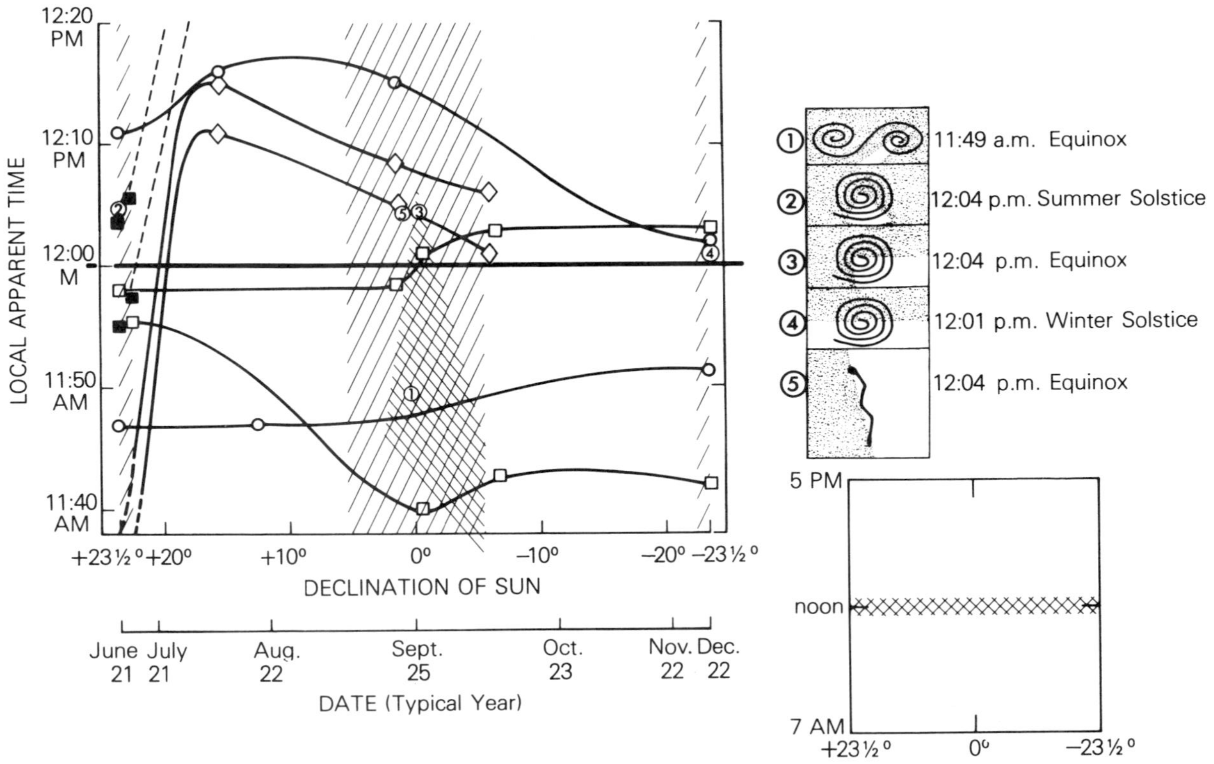

Figure 4.10. Left: Time that shadow first touches a glyph and then just covers it (bottom and top curves of each pair). East site: ■ rectangle, ◇ snake, ○ spiral. West site: □ rectangle, \\\\ range of marking on double spiral. Most points represent multiple observations on adjacent days. ///// are the intervals of one month centered on solstices and equinox.

Right: Moments near noon when shadow edges are centrally aligned on certain glyphs. These cases are located by numbers on the graph (left). (1) West site double spiral, March 20, 1984. East site spiral: (2) June 22, 1978; (3) September 20, 1979; (4) December 20, 1978. (5) Snake, September 20, 1979.

Inset: Replotting of graph on left to show the fraction of daylight hours involved in the markings. Illustration by Pat Kenny, © The Solstice Project especially when two symmetric shadow alignments occur simultaneously on two nearby glyphs. Fourth, six of the seven markings occur in such simultaneous pairs and are formed by unrelated shadows cast by different parts of the butte. Fifth, three markings use the same glyph in different patterns at different seasons. And sixth, four of the markings bisect or involve the centers of spiral glyphs (in a manner similar to markings at the three-slab site).

The rock edges casting the pattern on the rectangular figure at the west site in winter are vertical and about 3 m away; in summer they are horizontal and about 30 m distant. It is remarkable that a point was found where the shadow patterns cast by such a wide range of the butte's irregular surfaces would remain vertical throughout the year and cross the glyph consistently at noon.

The glyphs are so placed that no other shadow crossings occur near noon except those that form the markings; in general the glyphs are either steadily illuminated or shadowed at other times. The markings occur at the universally recognized extremes or midpoints in the annual solar cycle, as do those at the three-slab site. There are no equivalent markings at these sites that occur at times other than the key points in the solar cycle.

The astronomical markings involve a significant fraction of the known rock art on Fajada Butte. There are about 20 clearly formed glyphs at various locations on the butte, primarily scattered over its uppermost one-third (Sofaer and Crotty 1977), and some 20 other amorphous peckings and scratches and historical graffiti. When the five glyphs that are involved in the east and west site noon/seasonal markings are added to the two glyphs at the three-slab site, the marked glyphs constitute about one-third of the Fajada Butte rock art. (The other two-thirds appear not to be astronomically marked.) At each site the marked glyphs are dominant in clarity, size, and number over the few other nearby ones. Four of the seven spiral glyphs on the butte are used in astronomical markings, as are the only two rectangular figures.

Although the markings are referred to here as occurring near noon, it is likely that they were designed to commemorate both a spatial and a temporal halving of the day by recording when the sun is due south. (Note that it is more accurate to determine meridian passage or noon by observing the sun's changing azimuth around a north-south line than by monitoring changes in altitude.)

Central to these accurate noon/seasonal markings are several important astronomical concepts and their implementation. First, the creators of the markings singled out for emphasis midday and the year's quarter points. Then they had to establish the north-south meridian to the necessary accuracy, note the sun's transit, and identify the specific times of the equinoxes and the solstices. With this information they examined the shadow transitions on the Fajada cliffs throughout the year for possible sites. The weathering of the sandstone edges that form the markings has made it difficult to determine whether the developers of the markings worked these edges or used them in their natural state.

We conclude that the three Fajada sites were made to exhibit the extremes and midpoints of basic astronomical cycles considered significant by the Chacoan people. The markings occur only in a narrow band of times of day and year that is centered on astronomically recognizable events. An argument could be made that the petroglyphs were originally placed with no regard for their illumination and that shadows could randomly cross some of them at these times (particularly when the sun and shadows are moving fastest). This argument, however, cannot easily explain the pairings of markings, the multiple use of certain carvings, the repetitions of the alignment of two simple shadow patterns with the basic geometry of the petroglyphs at just the times of astronomical significance, or the consistent use of the spiral motif.

FURTHER OBSERVATIONS AND ANALYSIS OF THE THREE-SLAB SITE

Lunar Markings

New evidence and further analyses of this site support earlier findings that the northern minor and major lunar standstills are marked at moon rising and refute a speculation that the moon's meridian passage is marked.

The large spiral glyph at the three-slab site contains a total of nine and one-half turns, making it unique among the numerous recorded rock art figures in Chaco Canyon. No other spirals recorded in the canyon have so many turns; most have no more than half as many. This fact strengthens the suggestion made earlier (Sofaer et al. 1982a:176) that this number was chosen for the spiral to represent the number of years in the lunar standstill cycle. During the period of 9-10 years between the minor and major standstills, the position of the shadow cast by the most northerly rising moon each year shifts gradually across the nine and one-half turns of the spiral.

Further examination of the site has disclosed a pecked groove tangential to the large spiral and aligned with the major standstill shadow marking. This groove is similar to the pecked groove aligned with the minor standstill shadow marking (Figure 4.3; Sofaer et al. 1982a:173); these alignments draw attention to the shadow patterns as markings (Figures 4.1, 4.2).

It has been reported (Sofaer et al. 1982a:178) that a standstill could potentially be marked at four moonrises in the years of the extremes. Further analysis shows that the criteria can be met by as many as 13 risings in some of these years (Michael Zeilik, personal communication 1981). This increased frequency would make the marking of the standstills that much easier to conceive of and achieve.

It was speculated earlier that a possible lunar marking could occur at the moon's meridian passage at the declination of the major standstill, when the moon achieves an altitude of about 82 degrees (Sofaer et al. 1979:290). Recent measurements at the site show that an overhanging cliff edge about 10 m above the slabs would block the rays of the moon at this altitude so that no such marking could occur.

It is likely that the ancient Chacoans noticed the 18.6-year lunar cycle in the course of making horizon observations. During the 9-10 years between the standstills there is at Chaco a 13-degree shift in the extreme northerly and southerly positions of moon risings and settings; this shift is centered around the winter and summer solstice positions. The lunar extreme positions remain close to each standstill limit for a year or more. (The lunar standstill cycle would also be evident as the moon stood higher or lower in the sky in successive years and in the correspondingly changing nocturnal shadow and light patterns in the canyon and around and in the buildings.) In Chaco, the open, flat horizons and the clarity of the desert air of a thousand years ago would have made such observations far from difficult, especially for a people skilled in surveying. These rhythms of lunar cycles would become quite evident after some years of observation. The creators of the Fajada lunar markings did not necessarily know the standstill cycle to any better accuracy than approximately a year, which is what can be perceived in the movements of the shadows on the large spiral of the three-slab site.

These lunar markings do not necessarily signify that the Chaco culture was predicting lunar eclipses. The 18.6-year cycle is of limited use for this purpose (Chapter 2, this volume). It is worth noting, in this regard, that the modern Pueblo cultures have shown far more interest in cycles than in unusual events (Ellis 1975:60-63). The extremes and midpositions of recurring cycles are themselves significant because they define the limits and order of the cosmos. Recognition of these cycles provides reason enough to commemorate them.

Interrelationships of the Markings and the Assembly of the Three-Slab Site

The three-slab site combines a number of markings in a unique manner. New analyses presented here provide insights into its design and development.

Because of the extensive seasonal altitude changes of the midday and morning sun and the angles of the slabs relative to the cliff face, many different portions of the slabs create different parts of the complex set of markings. Nine separate surfaces are involved in casting the shadows that form six markings of five different declinations. (Note a possible seventh marking: Light is almost entirely blocked by the left and middle slabs at summer solstice [Sofaer et al. 1979:286]; only 10 days later a streak of light is much more evident and then grows through the months to become the bisecting equinox pattern. The effect at summer solstice may have been intended to focus attention solely on the bisecting pattern on the large spiral [Table 4.1]).

Each of the five solar and lunar markings on the large spiral interlocks with the positions of the other markings, in that each of these markings falls on the points and lines of symmetry that define the large spiral (Figure 4.2). There are five points that define the spiral: center, left edge, right edge, top, and bottom. The first four of these points are clearly part of the astronomical patterns. At the extremes and midpositions of the solar and lunar cycles a pattern of shadow and light strikes one and sometimes two of them. Conversely, most of these four points are involved in two markings. To a lesser extent the fifth point, the bottom edge of the spiral, is also defined by the markings.

At summer solstice the bisecting dagger-shaped pattern is centered simultaneously on both the vertical and horizontal axes of the spiral (Figure 4.1). The minor lunar standstill marking, which is formed by a different rock edge, also crosses the center of the large spiral.

Three shadow markings are tangent to the right and left edges of the large spiral. These are the winter solstice pattern, the lunar major standstill marking, and the lunar/solar 0-degree declination marking (Figure 4.1). These markings emphasize the right and left edges of the spiral.

The top edge of the large spiral is defined by another aspect of the markings. At summer solstice the dagger-shaped pattern starts as a dot of light on the top turn of the spiral. Only 10 days earlier or later this effect is lost: the first light on the cliff face is 10 cm above the spiral. The bottom edge of the spiral is related to, and hence perhaps defined by, the momentary symmetrical positioning (mentioned above) of the summer solstice light dagger.

It is possible to arrive at some estimates of the degree of sensitivity in the relationships among the rock slabs and the markings. Because of the oblique angle between the slabs and the cliff face, moving the rock edge that casts the moonrise shadows 1 cm to the north or south would create twice this displacement in the positions of the lunar shadows on the spiral and thus discernibly shift the markings. A movement of either the eastern or middle slab by 2-3 cm to the east or west would displace the shadows that form the midday light patterns by that much. The summer solstice pattern is only 2 cm wide, so such a change could block the light entirely. Because each of the slabs forms a part of up to five markings, any such shift would change and probably destroy more than one marking.

A recent report (Newman et al. 1982) presents a scenario in which the rocks could have fallen naturally close to their present positions. The evidence presented in that report does not, however, exclude the possibility of later deliberate movement of the rocks to create the markings. Indications of possible shaping of the slabs and the cliff face where the spirals were pecked are discussed in an earlier publication (Sofaer et al. 1979:289).

Both the interlocking nature of the markings and their sensitivity to the exact positions and shapes of the rock slabs indicate that moving and shaping of the slabs very likely took place. A few people could have moved and adjusted the slabs in small increments by simple techniques. Such manipulation would have made it easier to attain the interrelated markings. The complexity of the markings suggests that extensive planning, observation, and experimentation were required to achieve them.

The technology for shaping, moving, and using large slabs was well established in Chaco Canyon (Hewett 1936:87-88; Mindeleff 1891:148) and at other places in this cultural area, and many of these slabs are significantly larger than those at the three-slab site (Hayden 1878:429; Mindeleff 1891:58, 147-148; Newcomb 1966:137). Some isolated and implanted slabs are also known to have been used in historical times for astronomical purposes (Mindeleff 1891:86, 148).

Caution is appropriate in assessing the degree of artificiality in structures of the Pueblo culture. The natural appearance of many historical and prehistoric sites, especially shrines, conceals elements of construction and may be intended for the purposes of protective secrecy and integration with nature (Alfonso Ortiz, personal communication 1983). Stevenson's description of a shrine at Zia Pueblo illustrates the subtlety of construction at these sites:

> [A] stone slab rested so naturally on the hillside that it had every appearance of having been placed there by other than human agency. The removal of the slab exposed two vases side

by side in a shallow cave [Stevenson 1894:90].

PREHISTORIC PARALLELS TO THE FAJADA MARKINGS

Shadow-light formations mark critical times in the solar cycle at many prehistoric sites in the Chaco cultural region and elsewhere in the world. Many share the motifs of the Fajada markings: daggerlike shapes of light, spirals, snakes, and vertical shafts channeling light patterns at noon.

It is reported that winter solstice is marked when the rising sun's rays are collimated by a window in Pueblo Bonito, a major structure in Chaco Canyon (Reyman 1976). At Hovenweep, in southeastern Utah, an ancient Pueblo community lying just to the northwest of the Chaco region, several light markings are reported (Williamson 1981:68-70). At the solstices and equinoxes light is collimated by portholes to fall in the corners and midpoints of tower structures. At a nearby rock art site two light daggers bisect two concentric circles and a spiral at summer solstice sunrise. Shadow/light formations mark these glyphs again at equinox sunrise, at which time they also mark a snakelike form. Recent reports of solstice and equinox markings at an eastern Arizona site near the Chaco region include a great number of bisecting dagger shapes of light on rock art, often on spirals (Chapter 11, this volume). Similar observations are reported in California (Krupp 1983:129-137).

A number of the sites of monumental architecture in Mesoamerica have been noted for their symbolic display of shadow and light (Aveni 1980:284-286). For example, a snake-shaped shadow is formed at equinox along the edge of a Mayan pyramid at Chichen Itza. While this marking is several hundredfold larger than the small snake marked by a shadow on Fajada's east site, the parallel is curious: both sites record equinox with shadow alignments on nearly vertical snakes. At two other sites in Mesoamerica the sun's rays at zenith passage are channeled through vertical shafts to fall as discs of light on the subterranean floors of the structures (Aveni 1980:253-256).

Despite these parallels the Fajada markings remain distinctive in at least three respects: their combining of noon and season, their combining of sun and moon, and their use of rock art and rock slabs in numerous markings clustered on a single, prominent butte. Some of the particular characteristics of the Fajada markings can be understood in the context of historical Pueblo culture.

ETHNOGRAPHIC BACKGROUND

Ethnographic accounts of the historical Pueblo cultures describe concepts fundamental to the Pueblo people's world view. These concepts often have significant parallels among findings about prehistoric Pueblo communities.

Some discontinuity between prehistoric and historical Pueblo culture over the intervening 800-1000 years should be expected as a result of environmental and cultural change and migration, including the Spanish entrada in the 1500s (Berry 1982; Cordell 1984; Upham 1984). Certain gaps exist in the ethnographic record of astronomical practices because the Pueblo traditions of secrecy protect all ceremonial activities, including solar and lunar observations.

> Everyone who has worked among Pueblo Indians realizes only

> too well how averse they are to revealing the details of this manner of life. This attitude on the part of native informants makes it virtually impossible to secure a complete record of any Pueblo tribe [Titiev 1944:4].

Fewer than 20 of the 80 or more historical pueblos extant when the Spaniards first came survived to be included in ethnographic studies. Thus, a great deal of information about Pueblo culture is lost or not available. The information that does exist can be used to provide general insights into the significance of prehistoric findings rather than to verify or refute interpretations of specific phenomena.

The historical Pueblos conceived of complementary roles of sun and moon (see references cited in Sofaer et al. 1982a) and used both of them intensively in timing of planting and ceremonial activity. These activities are consistent with the overlapping systems of marking of the solar and lunar cycles at the three-slab site. A recent study (Tedlock 1984:6) describes clearly the complementary roles of sun and moon at Zuni. In timing of the ceremonies for the solstices, the "weak light" of the winter solstice sun is matched with the "bright light" of the full moon, and the "bright light" of the summer solstice sun is matched with the "weak light" of the new moon. The location of the astronomical markings high on the seemingly remote and inaccessible Fajada Butte is in keeping with ethnographic reports of the historical cultures. Buttes, mesas, and mountains are often regarded as sacred places, and shrines are placed on their tops (Boas 1925-1928:39-40; Dumarest 1919:206-207; Ortiz 1969). Some shrines on high sites are used for solstice ceremonies (Kallestewa et al. 1984; Stevenson 1904:109, 149).

Specific features of the complexly organized communities of the prehistoric Pueblos are not found or reported among the historical Pueblos. For example, the elaborately planned multistory architecture and engineered road system of the Chaco culture are not present among the historical Pueblos. Precise equinox markings and cardinal alignments are also not evident. The general period of the equinoxes is of major significance among the Tewa Pueblos, however (Ortiz 1969). Similarly, there are no reports in the ethnographic record of knowledge or markings of the 18.6-year lunar cycle among the historical Pueblo cultures, although there is a possible hint of previous knowledge of this cycle in the cosmogony of one pueblo (Stevenson 1894:71).

There are numerous reports of Pueblo people observing shadow and light patterns cast by upright slabs or by windows and doorways on walls and floors of houses, kivas, and chief's houses to time ceremonial and agrarian activities (Cushing 1979:117; Lange 1959:56, 249; Mindeleff 1891:86, 148). On ceremonial occasions light might fall on a quartz crystal, a bowl of water, a deer skin, or a kiva bench (Lowie n.d.; Parsons 1929:176; Titiev 1944:105).

Meridian passage is known to be significant in the cosmology of historical Pueblo people and is an integral part of many myths and ceremonies, often involving light markings. One such instance involves the winter solstice noon.

> In the roof of the ceremonial room there is a hole through which at noon the sun shines on a spot on the floor near where the chief stands. . . . All

> sing the song of "pulling down the sun." . . . This is noon time when for a little while the Sun stands still [Parsons 1932:292-293].

A similar ritual is practiced at summer solstice (Parsons 1932:297). In keeping with the complementary roles of sun and moon, a report states, "The ritual of bringing down the moon seems to be much the same as that of bringing down the sun," and in this ritual the moon is said to stay "until noon" (Parsons 1932:330). Another study reports that the moon's meridian passage is noted in the timing of Pueblo ceremony (Tedlock 1984:94, 108).

Several other ritual practices and traditions of the Pueblo culture convey the significance of noon as the time when the sun stands still or rests for a short time in the middle of its course through the day (Boas 1925-1928:284; Dumarest 1919:222) and as the time when "he stops for dinner" (Curtis 1926:104; Dumarest 1919:222). The cacique conducts ceremonies in which the sun as a disc is moved across a screen in the arc of its day's course; he orders the sun at the middle of its course to stop for a short while at its noon position (Dumarest 1919:198; Lange 1959:267).

In one pan-Puebloan tradition, the Sun impregnates a virgin and she gives birth to the son(s) of the Sun (Parmentier 1979:609). The impregnation and the birth often occur at noon and sometimes near summer solstice (Benedict 1931:31, 1935:46; Cushing 1931:429, 431, 436; Dumarest 1919:217; Gunn 1917:129; Parsons 1932:393, 1940:55-56; Stephen 1929:11-14). In several versions of this tradition the sun's rays penetrate a window or hatchway of a dwelling and fall on the lap of the maiden. In one version the virgin (yellow woman of the north) stops to rest on her journey to the center of the earth, and the Sun when it is "over the middle of the world" at the "middle of the day" embraces her and she becomes pregnant (Stevenson 1894:44-45). This theme of the impregnating power of the sun's rays is also evident in the tale of the Sun's son who, as a baby, identifies his father by crawling to the rays of the sun on the kiva floor (Alfonso Ortiz, personal communication 1981; Parsons 1940:56-57).

The use of the rattlesnake as a marked glyph is consistent with the snake's wide ceremonial use and symbolic meaning in historical Pueblo culture, including a close association with the sun. A rattlesnake effigy and a ceremonial staff with the rattles of a rattlesnake indicated on it were found at Pueblo Bonito, which suggest a possible ritual involvement with the rattlesnake as part of the Chaco culture (Pepper 1920:147). Among its many roles, the snake connects the below and above worlds (Tyler 1964:222, 234). At one pueblo it is connected with the zenith and the sun: "Huwaka (Serpent of the Heavens) has a body like a crystal, and it is so brilliant that one's eyes cannot rest upon him; he is very closely allied to the sun" (Stevenson 1894:69). Similarly, at another pueblo a snake with glistening scales is said to fly up to the sun each day and brilliantly reflect the sun's rays (Charles Loloma, a Hopi religious leader, personal communication 1983). Some reports on Pueblo cultures also connect the snake with fertility and equinox (Tyler 1964:228-229, 245-247).

While there is no conclusive information concerning interpretations of the spiral among the historical Pueblos, some Pueblo people have indicated that it conveys the movement of people and clans. In one report, the spiral was described as the movement of people in their search for the center of the earth in the origin story (Roberts 1932:151). The spiral has also been reported to represent the annual movement of the sun (Charles Loloma, personal communication 1983).

ASSOCIATIONS WITH THE PREHISTORIC PUEBLO CULTURE AND CHACO SOCIETY

The Fajada markings were probably developed when Pueblo people settled and flourished in the canyon between about AD 900 and 1300. The spiral petroglyph is identified with prehistoric Pueblo people in this region during this period (Schaafsma 1980:135-136; Polly Schaafsma, personal communication 1983).

The achievements of the Chaco culture--elaborate roads, irrigation works, and multistory architecture, all constructed between AD 1000 and 1150--involved highly developed skills of planning, engineering, and surveying (Kincaid 1983; Lekson 1984; Marshall et al. 1979; Powers et al. 1983; Vivian 1974; also see Chapter 1, this volume). During this period the Chaco society used these skills to align accurately several major constructions to the cardinal directions. Primary elements in the symmetric designs of the isolated great kiva at Casa Rinconada and of one of the central pueblos, Pueblo Bonito, and the initial portion of the North Road are oriented to within 0.25-0.5 degrees of the cardinal points (Sofaer et al. 1986; Stein 1983:8-1; Williamson et al. 1977:208-212; the canyon alignments were confirmed by surveys carried out by the Solstice Project).

The north/south alignments are particularly significant to the phenomenon of noon markings. A likely method of identifying noon is to watch shadows from a vertical object cross the north/south line as the sun crosses the meridian. Knowledge of true north within 0.25 degrees permits knowledge of solar noon at the latitude of Chaco to within less than one minute throughout the year. Similarly, knowledge of east and west to within less than 0.5 degrees allows the determination of the day of equinox. The architectural alignments indicate the Chaco culture's interest in and capability of achieving such accuracy. These alignments may also represent divisions of space that correspond to the markings on Fajada that divide the day and the year.

CONCLUSION

Although the astronomical markings and alignments of Chaco incorporate utilitarian calendric information, they do so with a redundancy and accuracy far beyond the practical requirements of time-keeping devices. For example, precise noon markings high on a steep butte serve no apparent useful purpose, nor do the markings of the lunar standstill cycle. The accurate alignments that define the major axes of the central ceremonial structures of the canyon are similarly abstract; for example, the east-west alignments could not have been used to determine (or have been determined by) equinox sunrise/sunset because of the locally elevated horizons. Similarly, the North Road was built elaborately and accurately in a direction that serves no apparent utilitarian purpose (Sofaer et al. 1986; Stein 1983:8-1). Rather, what may be seen in the alignments and markings is the geometric expression of astronomical concepts and of the culture's cosmology.

Chaco Canyon was the center of an extensive road network and outlier system. Recent analysis indicates that it may have been a center for pilgrimage and ritual (Chapter 1). Fajada Butte may have been a center for the culture's ritual activity related to the sun and the moon.

The clustering of markings found there is so far unique; surveys by the Solstice Project of all the prominent landforms in the Chaco cultural region have disclosed no further astronomical marking sites. When the sun in "the middle of the day" is over "the middle of the earth," the butte's glyphs commemorate this special moment in time and space. The extremes and midpoints of the solar and lunar cycles are integrated on Fajada. The daily and seasonal passages of the sun are united, as are the sun and moon and the earth and sky in the play of shadow and light on the rock carvings atop Fajada Butte.

ACKNOWLEDGMENTS

We are indebted again to Walter Herriman, former superintendent, and the staff of the Chaco Culture National Historic Park for their continued help and cooperation. Peggy Wier, Michael Marshall, Karl Kernberger, James Grant, Colin Franklin, Kenneth Butterfield, and Sheila Rotner helped collect the data, often under adverse conditions. We are grateful to Alfonso Ortiz, Gerald Hawkins, Fred Eggan, Cesare Marino, Rob Blair, Evelyn Newman, Robert Mark, Harold Malde, Fred Nials, and particularly LeRoy Doggett and Stephen McCluskey for a number of helpful conversations. Pat Kenny is to be especially thanked for preparing the illustrations. We again thank our families and friends for their assistance and understanding.

[Note: We understand that an accompanying paper (Chapter 5, this volume) offers other explanations for some of the phenomena discussed here. We have not been given an opportunity to review this alternative analysis or to comment on it.]

REFERENCES CITED

Aveni, Anthony
1980 Skywatchers of Ancient Mexico. University of Texas Press, Austin.

Benedict, Ruth Fulton
1931 Tales of the Cochiti Indians. Bureau of American Ethnology Bulletin 98. Washington, D. C.
1935 Zuni Mythology. Columbia University Contributions to Anthropology 21. New York.

Berry, Michael S.
1982 Time, Space and Transition in Anasazi Prehistory. University of Utah Press, Salt Lake City.

Boas, Franz
1925-1928 Keresan Texts (2 parts). Publications of the American Ethnological Society 8. New York.

Cordell, Linda S.
1984 Prehistory of the Southwest. Academic Press, New York.

Curtis, Edward S.
1926 The North American Indian 17. The Plimpton Press, Norwood, Massachusetts.

Cushing, Frank M., compiler and translator
1931 Zuni Folk Tales (reprint). Alfred A. Knopf, New York. Originally published in 1901 by G. P. Putnam's Sons, New York.
1979 Zuni (reprint), edited by Jesse Green. University of Nebraska Press, Lincoln. Originally published in 1882 by The Peripatetic Press, Santa Fe.

Dumarest, Noel
1919 Notes on Cochiti, New Mexico. Memoirs of the American Anthropological Association 6(3), edited by Elsie C. Parsons. Lancaster, Pa.

Ellis, Florence H.
1975 A Thousand Years of the Pueblo Sun-Moon-Star Calendar. In Archaeoastronomy in Precolumbian America, edited by A. F. Aveni, pp. 58-87. University of Texas Press, Austin.

Gunn, John M.
1917 Schat-chen: History, Traditions and Narratives of the Queres Indians of Laguna and Acoma. Albright and Anderson, Albuquerque.

Hayden, F. V., editor
1878 Tenth Annual Report--Geological and Geographical Survey of the Territories. Washington, D. C.

Hewett, Edgar L.
1936 The Chaco Canyon and Its Monuments. University of New Mexico Press, Albuquerque.

Kallestewa, B., J. Niiha, A. Peywa, A. Pinto, R. Quam, A. Nastacio, and W. Eriacho
1984 Statements by Zuni Religious Leaders on Kolhu wala:wa. Testimony presented to the Select Committee on Indian Affairs, U.S. Senate, April 3, 1984. Washington, D. C.

Kincaid, Chris, editor
1983 Chaco Roads Project Phase I. Bureau of Land Management, Albuquerque.

Krupp, E. C.
1977 In Search of Ancient Astronomies. Doubleday, Garden City, New York.
1983 Echoes of the Ancient Skies. Harper and Row, New York.

Lange, Charles M.
1959 Cochiti: A New Mexico Pueblo, Past and Present. University of Texas Press, Austin.

Lekson, Stephen
1984 Great Pueblo Architecture of Chaco Canyon, New Mexico. National Park Service Publications in Archeology 18B. Albuquerque.

Lowie, R. H.
n.d. "Hopi Indian Inside a Kiva." Photo No. 283545. American Museum of Natural History, New York.

Marshall, Michael P., John Stein, Richard Loose, and Judith E. Novotny
1979 Anasazi Communities of the San Juan Basin. Public Service Company of New Mexico, Albuquerque, and New Mexico State Historic Preservation Bureau, Santa Fe.

Mindeleff, Victor
1891 A Study of Pueblo Architecture. In Eighth Annual Report of the Bureau of American Ethnology, 1886-1887 pp. 3-228. Washington, D. C.

Newcomb, F. J.
1966 Navajo Neighbors. University of Oklahoma Press, Norman.

Newman, E. B., R. K. Mark, and R. G. Vivian
1982 Anasazi Solar Marker: The Use of a Natural Rockfall. Science 217: 1036-1038.

Ortiz, Alfonso
1969 The Tewa World: Space, Time, Being, and Becoming in a Pueblo Society. University of Chicago Press, Chicago.

Parmentier, Richard J.
1979 The Pueblo Mythological Triangle: Poseyemu, Montezuma, and Jesus in the Pueblos. In Southwest, edited by Alfonso Ortiz, pp. 609-622. Handbook of North American Indians, Vol. 9. Smithsonian Institution, Washington, D. C.

Parsons, Elsie Clews
1929 The Social Organization of the Tewa of New Mexico. Memoirs of the American Anthropological Association 36. Menasha, Wisconsin.
1932 Isleta, New Mexico. In Forty-Seventh Annual Report of the Bureau of American Ethnology for the Years 1929-1930, pp. 193-466. Washington, D. C.
1940 Taos Tales. Memoirs of the American Folklore Society 34. G. F. Stechert, New York.

Pepper, George M.
1920 Pueblo Bonito. Anthropological Papers of the American Museum of Natural History 27. New York.

Powers, R. P., W. B. Gillespie, and S. H. Lekson
1983 The Outlier Survey. Reports of the Chaco Center 3. National Park Service, Division of Cultural Research, Albuquerque.

Reyman, Jonathan E.
1976 Astronomy, Architecture, and Adaptation at Pueblo Bonito. Science 193:957-962.

Roberts, F. H. H., Jr.
1932 The Village of the Great Kivas on the Zuni Reservation, New Mexico. Bureau of American Ethnology Bulletin 111. Washington, D. C.

Schaafsma, Polly
1980 Indian Rock Art of the Southwest. University of New Mexico Press, Albuquerque.

Sofaer, A., and J. Crotty
1977 Survey of Rock Art on Fajada Butte. Ms. on file, Chaco Center, University of New Mexico, Albuquerque.

Sofaer, A., M. P. Marshall, and R. M. Sinclair
1986 Cosmographic Expression in the Road System of the Chaco Culture of Northwestern New Mexico. Paper presented at the Second Oxford Conference on Archaeoastronomy, Merida (Yucatan), Mexico.

Sofaer, A., R. M. Sinclair, and L. E. Doggett
1982a Lunar Markings on Fajada Butte. In Archaeoastronomy in the New World, edited by A. Aveni, pp. 169-181. Cambridge University Press, Cambridge and New York.
1982b Noon Markings on Fajada Butte, Chaco Canyon, New Mexico. Bulletin of the American Astronomical Society 14:872.

Sofaer, A. V. Zinser, and R. M. Sinclair
1979 A Unique Solar Marking Construct. Science 206:283-291.

Stein, John R.
1983 Road Corridor Descriptions. In Chaco Roads Project Phase I, edited by C. Kincaid, pp. 8-1 - 8-15. Bureau of Land Management, Albuquerque.

Stephen, A. M.
1929 Hopi Tales. Journal of American Folklore 42:1-72.

Stevenson, Matilda Coxe
1894 The Sia. In Eleventh Annual Report of the Bureau of American Ethnology for the Years 1889-1890, pp. 3-157. Washington, D. C.
1904 The Zuni Indians: Their Mythology, Esoteric Fraternities, and Ceremonies. In Twenty-Third Annual Report of the Bureau of American Ethnology for the Years 1901-1902, pp. 3-634. Washington, D. C.

Tedlock, Barbara
1984 Zuni Sacred Theater. American Indian Quarterly 7(1):93-110.

Titiev, Mischa
1944 Old Oraibi: A Study of the Hopi Indians of Third Mesa. Papers of the Peabody Museum of American Archaeology and Ethnology 22(1). Harvard University, Cambridge, Massachusetts.

Tyler, Hamilton A.
1964 Pueblo Gods and Myths. University of Oklahoma Press, Norman.

Upham, Steadman
1984 Polities and Power--An Economic and Political History of the Western Pueblo. Academic Press, New York.

Vivian, R. Gwinn
1974 Conservation and Diversion: Water-Control Systems in the Anasazi Southwest. In Irrigation's Impact on Society, edited by Theodore Dawning and McGuire Gibson, pp. 95-112. Anthropological Papers of the University of Arizona 25. Tucson.

Williamson, Ray A.
1981 North America: A Multiplicity of Astronomies. In _Archaeoastronomy in the Americas_, edited by R. A. Williamson, pp. 61-80. Ballena Press, Los Altos, California.

Williamson, R. A., H. J. Fisher, and D. O'Flynn
1977 Anasazi Solar Observatories. In _Native American Astronomy_, edited by A. F. Aveni, pp. 203-217. University of Texas Press, Austin.

Chapter 5

Romancing the Stone, or Moonshine on the Sun Dagger

John B. Carlson

In June of 1977, artist Anna Sofaer was a volunteer participant in a rock art recording field school in Chaco Canyon, New Mexico. She was given the famous 135 m high Fajada Butte at the south entrance of the canyon to explore as part of the summer survey. According to published accounts (e.g., Frazier 1978, 1979; Marshall 1982; Sofaer et al. 1979a, 1979b), it was just before midday on June 29 at an obscure petroglyph site near the top of the butte that Sofaer serendipitously first observed the light-and-shadow phenomenon that has come to be known as the "Sun Dagger" (Sofaer and Ihde 1982).

The petroglyph site consists of two spiral-type petroglyphs pecked into the cliff face of the butte and situated behind three near-vertical 2-3 m high sandstone slabs balanced precariously on soil and small stone fragments (Figure 5.1). Sofaer and archaeologist Jay Crotty watched as light shining between Slabs 1 and 2 (Sofaer et al. 1979b:284-285) formed a spot at the top of the large spiral, lengthened to create a vertical dagger shape of light spanning the petroglyph, and then moved downward rapidly to disappear near the bottom of the spiral. The whole event, which has now been well documented with time-lapse photography, takes approximately 18 minutes from start to finish. The effect is striking; observing the time-lapse sequence has convinced most researchers that the large spiral petroglyph was intentionally positioned by its maker to interact in some way with the "sun dagger" phenomenon.

Because of the striking appearance of the dagger of light, the fact that it virtually bisected the spiral petroglyph, and because it was June 29, Sofaer was compelled to believe that the phenomenon was created intentionally to mark the summer solstice specifically--that it was not simply a daily pattern that repeated essentially unchanged throughout the year. As a result of many subsequent observations of the site by Sofaer and others, her initial inspiration that the dagger of light would be found to be centered on the petroglyph around the time of the June solstice has been adequately verified. The petroglyph site serves and probably was created to serve as at least an approximate marker of the summer solstice. The creators were suggested plausibly, but not conclusively, to have been the Anasazi Indians who occupied Chaco Canyon from about AD 400 to 1300 (Reyman 1980; Sofaer et al. 1979b).

Figure 5.1a. View of the three-slab site from the south. The three near-vertical sandstone slabs are seen at lower right in the context of their location within 10 m of the summit of Fajada Butte. Photograph by J. Carlson.

Figure 5.1b. Close-up view of the three-slab site from the south. Slab 1 is to the right (east), Slab 2 is in the center, and Slab 3 is on the left. Note that a portion of a "sun dagger" created by light passing between Slabs 2 and 3 can be seen between Slabs 1 and 2. Photograph by J. Carlson.

Figure 5.1c. View of the three-slab site looking from the west. Considerable weathering of the exposed surfaces is evident. Photograph by J. Carlson.

Figure 5.1d. View of the three-slab site looking straight down from the overhanging edge of the mesa summit. Photograph by J. Carlson.

Figure 5.1e. View of the three-slab site from the east showing the natural lean-to. A "sun dagger" is seen on the cliff face at the approximate location of the large spiral petroglyph. Photograph by J. Carlson.

Figure 5.1f. View of the three-slab site from the east. Note the sharp fracture plane of the sunrise shadow-casting edge of Slab 1 and its parallel alignment with the inner faces of Slabs 2 and 3. Photograph by J. Carlson.

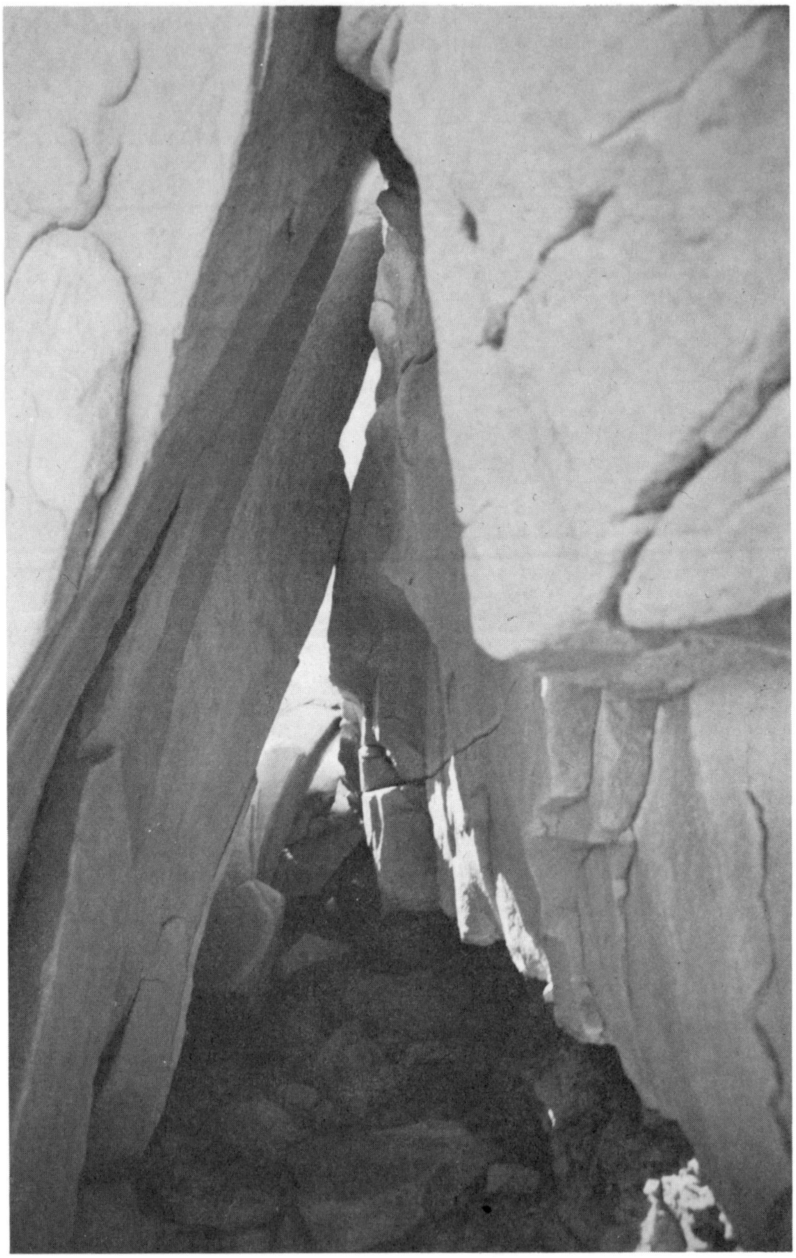

Figure 5.1g. View inside the natural lean-to of the three-slab site from the east. Note the crisp, parallel fracture planes of the three slabs and a portion of the sunrise shadow-casting edge of Slab 1. Photograph by J. Carlson.

If this had been the whole story--that the Fajada site contained a visually pleasing light-and-shadow-on-petroglyph marker of the summer solstice, possibly created by the Chacoan Anasazi--we would have simply yet another example of a Native American archaeoastronomical site involving plays of light and shadow on petroglyphs. Williamson and Young (1979) first presented a discussion of an Anasazi site at Hovenweep National Monument that subsequently (Williamson 1979) was shown to contain a compelling sun-dagger-type summer solstice marking. Other examples include those proposed by Preston and Preston (Chapter 11, this volume) for petroglyph sites in Arizona. Sofaer and her colleagues (Sofaer et al. 1979a, 1979b) propose a far more elaborate set of claims for the Fajada site, however. They claim that the site not only marks the summer solstice with a sun dagger that bisects the large spiral, but that two daggers would bracket this spiral on either side at the winter solstice and that the second, smaller spiral to the left would be bisected by a short shaft of light on the equinoxes. Furthermore, they claim to demonstrate that the site was a "construct." Unlike all other such petroglyph marker sites that have been proposed, this one is alleged to have been constructed by the Chacoans, who selected and moved the slabs into their near-vertical position and modified the shadow-producing edges (Figures 5.1-5.3). Finally, they note that "the patterns formed at night by moonlight shining between the slabs are just as clear and noticeable as those formed by the sun" and that "extrapolations from the solar data suggest that a significant marking of the maximum lunar declination may occur" (1979b:287).

In the concluding paragraphs on the "possible significance of the lunar markings," they state that (a) "The Indians of the Southwest knew of the 18.6-year cyclic extremes in lunar declination [the lunar standstills] and incorporated that knowledge in a major archaeological site [at Casa Grande]" and that (b) "Striking alterations of patterns would indicate the alignment of sun, earth, and moon that makes possible a midwinter or midsummer solstitial or an equinoctial eclipse." This knowledge of the 18.6-year lunar standstill cycle is suggested to be reflected in the 19 half-turns of the large spiral. All of these assertions are based on the flimsiest of evidence and include an apparent confusion of the 29.53-day period of the moon's phases, the subject of the well-known Pueblo interest in the lunation, with the far more subtle 18.6-year cycle of lunistices. To date, there is no convincing evidence that the peoples of native America knew of or recorded the lunistices and their cycle (e.g., see Aveni 1985). As discussed later, Sofaer and Sinclair have abandoned the specific lunar alignments involving the moon near meridian transit and offer a new set of moonrise shadow effects that allegedly mark the northern major and minor lunar standstills as well as moon (or sun) rise at zero degrees declination (Sofaer et al. 1982).

Beyond the initial suggestion that the three-slab site, as it is now called (Sofaer and Sinclair 1985), was a summer solstice marker, most of the proposed astronomical functions of the site are speculative and the subject of considerable scientific controversy. This controversy has been intensified by media coverage that presents the sun dagger and all of its complexity to the public as established scientific fact--"an American Stonehenge" constructed by "Chacoan Einsteins" as it were. Science writer Kendrick Frazier first promulgated the speculative "construct" hypothesis in

Science News (1978); a year later he created considerable new mythology in an article (Frazier 1979) for Science 80 with diagrams of a four-part seasonal calendar with neat daggers such as never actually appear at the site (see Figures 5.2 and 5.3).

The main event of this media coverage was "The Sun Dagger," an hour-long pseudodocumentary produced and directed by Sofaer and Ihde (1982), which was aired nationwide on public television. The story line graphically tells how artist Anna Sofaer, triumphing over scientific disbelievers, has demonstrated that not only does the site mark summer solstice, but that it also indicates the equinoxes and winter solstice, giving a full annual solar calendar. Furthermore, it is called a "construct." The three 1000 kg slabs of sandstone are said to have been moved into place by the Anasazi, who also aligned the stones to work as a lunar marking construct to indicate the 18.6-year cycle of major and minor lunar standstills.

The final triumph in the story came when Sofaer went to England to present her results before the September 1981 Oxford Archaeoastronomy Symposium, a unique gathering of European and American researchers. To quote from the film, "The conclusion reached by this international group of scholars was that Fajada Butte is presently the only known site in the world where the extreme positions of both the sun and the moon are marked." It is indicative of the film that it characterizes the suggestion that Stonehenge marked both extremes of the sun and moon as "controversial," the same claim is given the alleged stamp of approval by the Oxford participants for the Sun Dagger. In fact, no such endorsement was ever made.

Following the airing of "The Sun Dagger" there were numerous media excesses, such as the "rock clock" of Fajada Butte being billed as "an astronomical and geometrical marvel" that "ranks with the Pyramids and Stonehenge" (Marshall 1982). All this "hype" greatly obscures reality by creating a durable popular mythology and makes the process of careful scientific evaluation very difficult.

RETHINKING THE SUN DAGGER

Scientific evaluation of the data and of the many multifaceted claims made for the Fajada Butte "three-slab site" in both the professional and popular literature has now begun in earnest. Reyman (1980) was one of the first to question many of the ideas put forward by Sofaer et al. (1979a, 1979b) and Frazier (1979), particularly with regard to their treatment of the relevant archaeological and ethnographic data. Newman et al. (1982) have provided convincing data and analyses demonstrating that the site is a natural rockfall, not a man-made construct. In Chapter 2 of this volume, Aveni provides further evidence casting doubt on Native American knowledge or use of the lunistices and lunar standstill cycle. Carlson (1983, 1984) and Hadingham (1984) reviewed "The Sun Dagger," questioning the methods, conclusions, and strong antiscientific bias of the film.

Zeilik (1985) has performed the most comprehensive reevaluation of the Fajada Butte solar marker. He approached the site on the basis of site integrity; archaeological, geological, and ethnographic context; and usefulness for confirmatory and anticipatory solar observations. He concludes that the site does not function as an accurate solar calendar, where "accurate" is defined in the context of the historical Puebloan culture, and that if used at all intentionally by the Anasazi for its relationship to the sun, the

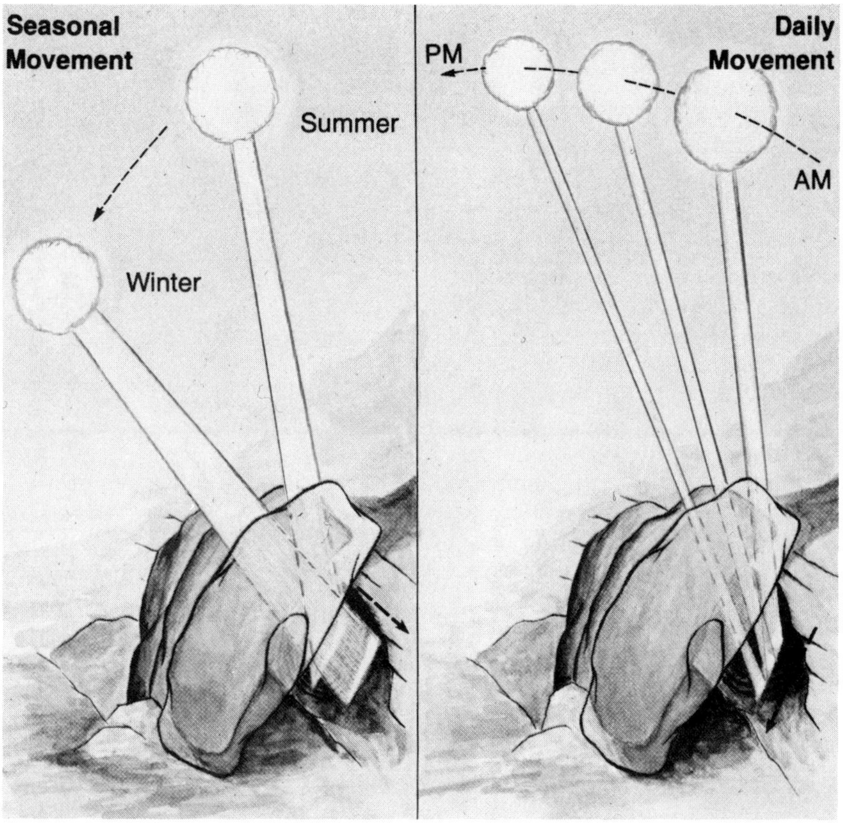

Figure 5.2. Diagram prepared by Mary Challinor for the Science 80 article by Kendrick Frazier (1979:61). "The slab's precise geometry transforms seasonal lowering sun into a rightward shift of light daggers on spirals, and the sun's daily horizontal passage into daggers that move down vertically."

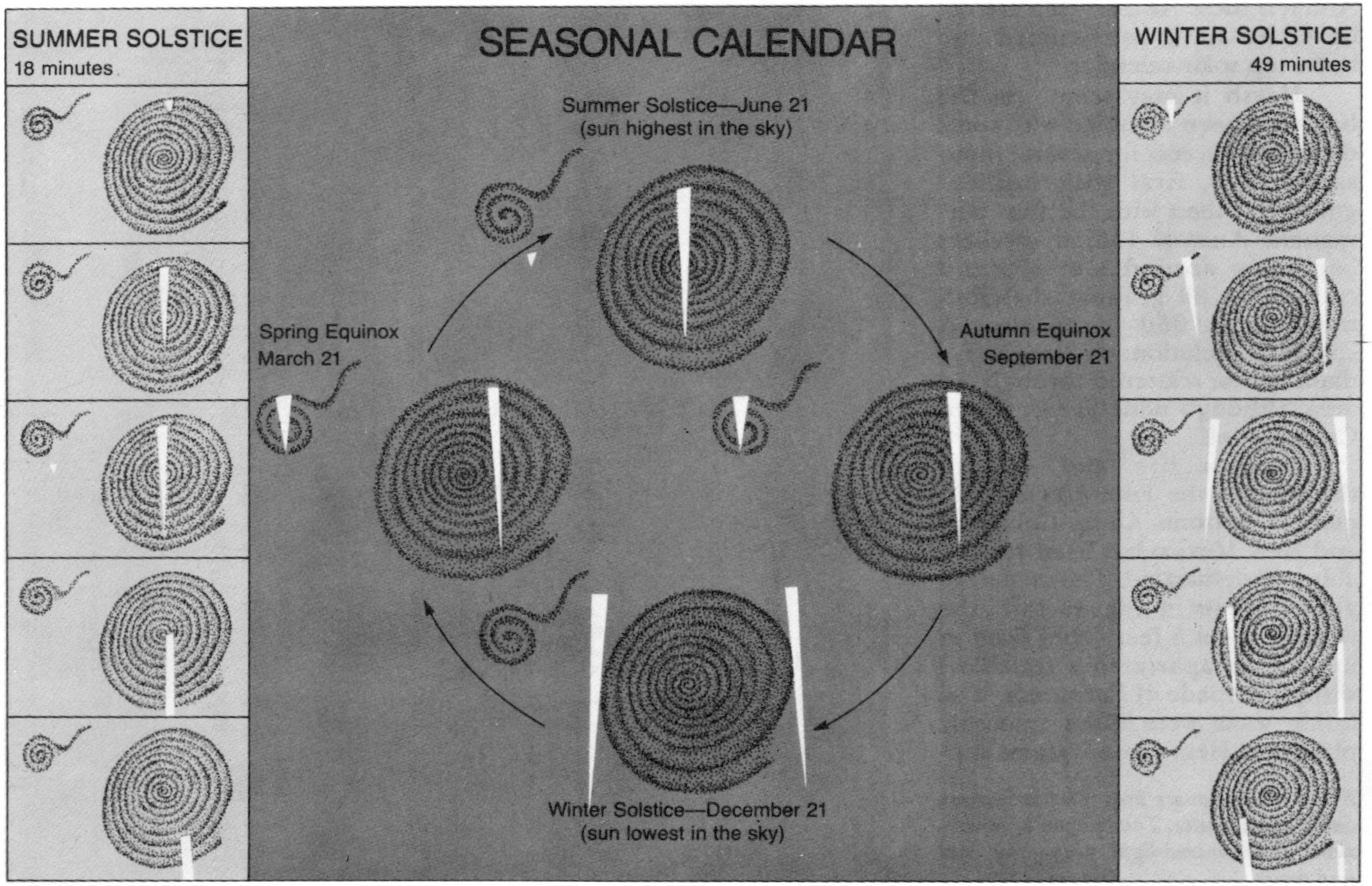

Figure 5.3. Schematic diagram of the hypothetical four-part "seasonal calendar" created by daggers of light falling on the two spiral petroglyphs at the three-slab site. Illustration prepared by Mary Challinor for the Science 80 article by Kendrick Frazier (1979:61), the stylized daggers are quite misleading in that they bear little resemblance to any of the light phenomena that actually appear at the site.

site most likely served as a shrine rather than as an observing station --a conclusion drawn from the characteristics of sun shrines among the historical Pueblos. "The place would be special and sacred, but not calendrical. It would act as a sacred showing of the sun--rather than an accurate calendar" (Zeilik 1985:S80). Furthermore, he evaluates the claim that the site also marks the northern declinations of the lunar standstill cycle and concludes that

> The case for lunar standstill alignments for the Fajada site fails to have support from ethnographic grounds. It also has no practical value as an adaptation for survival. Even if the site did mark the northern declinations of the major and minor standstills, it does so crudely and cannot be used to predict the year in which the standstill will occur. As an alternative interpretation, the site can be used at sunrise to provide a planting calendar, to anticipate the summer solstice, and to mark the equinoxes--functions much more in harmony with the ethnoastronomy of the historic Pueblos than observations of the lunar standstill cycle [Zeilik 1985:S84].

The present essay does not attempt anything like a comprehensive reevaluation of the three-slab site. Instead, I would like to rethink the full scenario of solar and lunar alignments proposed for the three-slab site as outlined by Sofaer et al. (1982) and more recently by Sofaer and Sinclair (Chapter 4, this volume). In particular, let us consider the three "solar markings at midday" and the three "solar/lunar markings at rising" illustrated in Figure 4.1 (previous chapter) in the context of the site as a natural rockfall, as a slightly modified natural rockfall, and as a substantial, engineered construction. This illustration is a corrected version of a diagram that appeared originally in Sofaer et al. (1982:Fig. 4).

THE FACTS AND THE CLAIMS

The specific set of hypotheses that we will take as given is the one outlined by Sofaer et al. (1982) and more recently by Sofaer and Sinclair (Chapter 4). Figures 4.1, 5.2, and 5.3 present this set in simplified diagrammatic form.

The three-slab site would never have been suggested as an archaeological site, man-made or otherwise, had it not been for the two spiral petroglyphs that are pecked into the vertical cliff face behind the slabs. The larger spiral to the right unwinds counterclockwise with approximately nine and one-half turns. It is elliptical and measures 34 by 41 cm. The smaller spiral to the upper left is actually a coiled snake with the head at the center. It has two and one-half turns, measures 9 by 13 cm, and uncoils from head to tail in a clockwise direction.

1. <u>Solar marking at midday</u> (+23.6 degrees declination). The vertical dagger formed when light passes between Slabs 1 and 2 (Figures 5.2 and 5.3) bisects the large spiral at the June solstice. In the proposed model, this determines the position of the center along the vertical axis. The top and bottom edges are fixed by the place where the initial spot of light first appears and roughly by the maximal hilt-to-tip dimensions of the dagger before it disappears. Though quite approximate, during the 18-minute sequence the location of the upper and lower boundaries and hence the center of the large spiral are delimited on the rock face.

Because the spiral is elliptical, its width is as yet undetermined.

2. Solar marking at midday (-23.6 degrees declination). Two vertical stripes formed by light passing between Slabs 1 and 2 and Slabs 2 and 3 "bracket the large spiral, holding it empty of light" (Sofaer et al. 1982:173). This winter solstice configuration thus operationally fixes the left and right edges of the large spiral. With the constraints imposed by the summer solstice display, the center and the horizontal and vertical boundaries have all now been determined. If the times of the solstices were already known by the usual horizon calendar methods, a sun watcher could in principle determine these parameters at the three-slab site in one year--given adequate observing conditions.

3. Solar marking at midday (0.0 degrees declination). A second dagger formed when light passes between Slabs 2 and 3 bisects the smaller spiral at the site. This dagger's vertical dimensions provide rough upper and lower boundaries for this spiral, thus determining a vertical and horizontal locus for the center. As an elliptical spiral, its width remains undetermined.

4. Solar/lunar markings at rising (0.0 degrees declination). At equinox sunrise the outer edge of Slab 1 casts a shadow over the far right groove of the large spiral (Figure 4.1). Of course, a similar shadow will be formed at moonrise when the moon is at zero degrees declination. This alignment is suggested to have been intentionally created. Note that the right edge of this spiral has already been determined by the winter solstice midday marker configuration.

5. Solar/lunar markings at rising (+18.4 degrees declination). Around the time of the minor lunar standstill, the shadow cast by moonlight grazing the outer edge of Slab 1 at the northern extreme moonrise covers the left half of the large spiral. This shadow passes roughly through the center of the spiral, and it is further alleged that the shadow is "aligned with a pecked groove (Sofaer et al. 1979b:Fig. 7) that runs from the spiral's center to the lower left edge, emphasizing this particular occurrence" (Sofaer et al. 1982:173-175). My own observations of the site have not convinced me that such an intentional groove actually exists. As noted above, the sun at sunrise will also produce such a shadow when it is at +18.4 degrees declination. Note that the center of the large spiral has already been determined by the summer solstice midday marker. Note also that such minor lunar standstills recur in an 18.61-year cycle.

6. Lunar markings at rising (+28.7 degrees declination). With calculations and simulation experiments, Sofaer et al. (1979b) show that around the time of major lunar standstill the shadow cast by moonlight grazing the outer edge of Slab 1 at the northern extreme moonrise falls roughly tangential to the far left edge of the large spiral (see Figure 4.1). The shift of the obliquity of the ecliptic for epoch AD 1000 was taken into account in their calculations, as were the effects of lunar parallax, atmospheric refraction, the lunar wobble of 9 arc minutes amplitude, and the local horizon height (Sofaer et al. 1982:176). As with the previously described lunar marking, Sofaer and Sinclair now report that there is a previously unnoticed pecked groove tangential to the large spiral and aligned with the major standstill shadow marking (Chapter 4). From my own inspections I have observed no such groove and reject this as valid data. Note that the dimensions and the left edge of the spiral have

already been determined by the winter and summer solstice midday marker configurations. As is the case for the minor standstills, the major lunar standstills recur in an 18.61-year cycle. Sofaer et al. (1982:176) further propose that the nine and one-half turns of the spiral may, as 19 grooves, record this 18.61-year lunar cycle or perhaps "are also a record of knowledge of the 19-year Metonic cycle." Here again there are hints of lunar eclipse prediction.

The initial motivations for a search for lunar standstill alignments at such an obviously solar site remain obscure. There do appear, however, to be two main sources of influence. The first is Stonehenge, with its complexity of solar and lunar standstill alignments and possibilities for lunar eclipse prediction, the forerunner of megalithic archaeoastronomical studies. Aveni (Chapter 2) has briefly discussed the history of this subject, noting how the paradigm has been uncritically transferred to other cultures and cults. The second influence has been noted by Sofaer et al. (1982:175): "A further stimulus to search for lunar significance within this assembly was provided by [Pueblo Indian anthropologist] Alfonso Ortiz (personal communication) . . ., who suggested that because of the dual roles of sun and moon in Pueblo culture, a site in which the sun was so clearly marked would also include the moon." This may or may not be true. When one looks at the record of known or suggested Pueblo sun-watching sites or sun shrines--be they ancient or contemporary, occurring in natural formations or in Pueblo architecture--one would be hard pressed to find a single example with alignments marking the moon's extremes as well as those of the sun. In any event, the suggestion that historical Pueblo moon watching involves alignments to the lunistices or knowledge of the 18.61-year standstill cycle is misguided. Any Southwestern ethnologist or student of the Pueblo cultures will attest to the symbolic and religious importance of the moon and the calendrically vital knowledge of the monthly lunation cycle, but at present there are no reliable data attesting to an awareness or cultural use of the lunistices and their cycle among the historical or ancient Pueblos.

Failing to discover and verify plausible lunar standstill markings produced near the moon's meridian passage, Sofaer and Sinclair (Chapter 5) sought and subsequently found moon<u>rise</u> alignments. The question is, are these alignments intentional or merely fortuitous? Corroboration must wait for further indisputable archaeological examples and solid ethnographic data. For now, we may gain insights by exploring the logic of the claims made for the three-slab site.

The Three-Slab Site: A Natural Rockfall

Despite the claims that the three-slab site was constructed, the best geological and archaeological data now strongly suggest that it is a natural rockfall. A series of facts support this. The three main slabs in their present relative positions form a "sandwich" that fell essentially intact from horizontal strata above and to the left of where the assemblage now lies. This has been documented in an article by two geologists and an archaeologist (Newman et al. 1982). They have shown that such natural configurations are common in Chaco Canyon and that none of the characteristic signs of Chacoan Anasazi masonry or other cultural remains are found at the three-slab site. The slabs are precariously balanced amid their own rubble and earth, with no foundation or buttressing to stabilize them (Figure 5.1). The outer exposed edges are heavily weathered, whereas the sides and protected inner edges appear to be as crisply fractured as the day they split from the cliff face. This can be seen clearly in Figure 5.1 or in the photograph provided by Sofaer et al. (1979b:cover). With these thoughts in mind, let us explore the ramifications for the program of six hypothetical solar and lunar alignments outlined above.

Obviously, if we accept that the three-slab site is entirely natural and that it functions now virtually as it did at the time of its creation (i.e., with a negligible minimum of erosion and subsidence), then all of the light-and-shadow engineering involved only the specific design and placement of the two spiral petroglyphs behind the slabs. The daggers of light are then purely natural, serendipitous visual phenomena, striking though they may be. If this were the case, the situation would be in accord with all of the comparable Native American rock art sites investigated thus far.

As we have seen, the three solar markings at midday at the solstices and equinoxes uniquely fix the locations and dimensions of the two spirals. For this reason the solar/lunar shadow markings at rising must be fortuitous and not created intentionally by design. Alternatively, if one were to argue that the rise shadow markings were designed, then the midday solstice markings must be serendipitous. This latter proposition seems to me to be most unlikely. In any case, if the rockfall was unaltered, it is impossible for both sets of markings to have been created by design. Given the lack of evidence for Pueblo lunistice watching, I reject the intentional creation of lunar markings at rising as an untenable hypothesis for this first scenario.

The Three-Slab Site: A Modified Natural Rockfall

It is well within the realm of possibility that the carvers of the two petroglyphs also slightly modified the natural rockfall and enhanced the serendipitous light phenomena that originally attracted their attention. The inner edges and mutually facing sides of the slabs are unmodified, and the tops and outside edges are so weathered that it is not possible to prove that any modification occurred (Figure 5.1). A close inspection of the three-slab and other similar sites will show that such daggers of light must necessarily form in vertically aligned sandwiches of rock that have fractured along flat sedimentary planes. In my view, modification of the light-dagger-forming surfaces may have occurred, but probably only by natural weathering. In any case, modification would have only a small effect on the essential visual outcome.

The real challenge relative to the second scenario would have been to make the three lunar markings at rising function in consort with the solstice markings at midday when these latter alignments had already uniquely fixed the location of the large spiral. If the rise alignments are not purely serendipitous, the Chacoan designer would probably have needed to make two types of modifications of the shadow-casting outer edge of Slab 1. First, this edge must have been modified and/or Slab 1 must have been levered in or out, toward or away from the cliff face, to align the shadow edge. Second, the projected distance from the shadow-casting edge to the large spiral would have to have been adjusted by moving Slab 1 toward or away from the spiral to adjust the angles subtended as the moonrise moved from 0.0 to +18.4 to +28.7 degrees declination. Since any movement of Slab 1 would disrupt the delicate geometry already established to mark the solstices at midday, only modification of the outer shadow-casting edge of Slab 1 would avoid this problem.

At this point, two relevant observations become important. Even a casual inspection of the shadow-casting edge of Slab 1 shows that it has not been modified by abrasion, grinding, or further fracturing. Figures 5.1e-g and the cover photograph published in Sofaer et al. (1979b) all support this conclusion. The full fracture planes on the insides of all three slabs are in virtually the same condition that they were when the slabs broke away from the point on the cliff face where they match the parent strata. Furthermore, as can also be seen in these photos (which are no substitute for on-site inspection), the inside fracture planes of the three slabs parallel each other rather closely. Slab 1 rests against the cliff face at roughly the same angle as the other two slabs; if Slab 1 was levered in or out, the others were moved in the same way and to essentially the same degree.

My view of the scenario of minor modifications of a natural rockfall is that the evidence argues strongly against either of the two types of alteration of the moonrise shadow-casting edge that would have been necessary to create these alignments. Hence, they are serendipitous at best and were unintended by and probably unknown to the carvers of the petroglyphs. Thus, I do not accept the alleged lunar significance of the nine and one-half turns of the large spiral.

The Three-Slab Site: A Substantial Engineered Construction

We know from their massive feats of masonry construction in the many ancient pueblos of Chaco Canyon that the Anasazi and later Pueblo peoples were capable of great engineering achievements in stone. For example, thick, one-ton, worked stone disks were placed as foundation supports for the great tree-sized posts that supported the roof beams in the great kivas at Casa Rinconada and Chetro Ketl. The three-slab site shows no evidence of the skilled handiwork of the Anasazi stonemason. Indeed, the 1000 kg slabs are precariously balanced, showing no evidence of foundation or buttressing--Anasazi or otherwise. The subtle, "natural" appearance of the stone was described by mythologist Joseph Campbell in "The Sun Dagger" film as being "somehow very Indian." The thought is good, but in this case the superior hypothesis, following Newman et al. (1982), is that the site looks so natural because it <u>is</u> natural, in contrast to the appearance of most other Pueblo sun shrines. If we nonetheless hypothesize that the three-slab site was a construction designed to do everything that has been claimed for it, we must see how this hypothesis fits the facts.

As discussed in detail above, the slabs are found in the same three-layered sandwich configuration that they exhibited when they formed part of the cliff, the facing sides and inner sides are nearly pristine (unweathered and unmodified), and the slabs are precariously balanced against the cliff face with no archaeological evidence for human use save for the two spiral petroglyphs. As also mentioned previously, no one would have suggested that the site was anything other than a natural rockfall had the sun dagger phenomenon not been discovered.

In spite of the seeming freedom to create available to the designers in this third scenario, in fact they would have had a much greater burden. They had somehow to visualize in advance the whole complex creation involving the erection of an existing ensemble of three sandstone slabs, presumably then lying flat below the cliff face amid additional rubble. Whatever they did, their end result would have to maintain this sandwich configuration with the inside edges unmodified because this is the way we find it today. As weathering is evident only on the currently exposed outer surfaces, the implication is clear that the ancient engineers found the slabs largely unweathered; all of the external weathering has to have taken place in the years since the hypothesized human creation of the site. Furthermore, the designer/engineers must have had a very good knowledge of the solar and lunar excursions along the horizon and near meridian transit as viewed from this site, and they must have accurately known the lengths of the annual solar solstice and almost 19-year lunar standstill cycles. To create this unique site today, both in conception and in execution, would be a formidable task for a modern astronomer/engineer, let alone a Pueblo sun priest. Yet the site does exist, and it apparently will "work" in a manner close to the way described by Sofaer et al. (1982).

My view is that, even given the freedom of design implied by this third scenario and the engineering capability for execution of this design, the physical and geometrical constraints as presented make the task almost impossible. I would be the last person to argue against the philosophy that "anything is possible under the sun," and perhaps under the moon, but I reject this third scenario as being far less likely than the second, though it seems at first glance to provide greater creative freedom for the Anasazi designer. At least in the second scenario, the essentials of the design already existed naturally, waiting to be discovered.

CONCLUSIONS

In rethinking interpretations of the three-slab site as comprehensively as possible, I conclude that it is a sun shrine marking the summer solstice and probably the winter solstice and equinoxes. The two spiral petroglyphs were carved behind a natural rockfall by the Chacoan sun watchers to mark these events, taking advantage of the natural patterns of light that have come to be called sun daggers. Any modification of the outer edges of the slabs would have little effect on the essential light phenomena. Crudely executed though they are, the spiral petroglyphs in conjunction with the natural lean-to formed by the slabs create an elegant and striking display of light marking the annual solar events. Previous knowledge of the times of the solstices and, in particular, of the equinoxes would have been necessary for the petroglyph carvers--knowledge certainly derived through the standard Pueblo practices of horizon sun watching. Despite claims to the contrary, no particular precision is manifest in the marking of the solar events at the three-slab site. The design of the site and its inaccessible location can be used to support the argument (Zeilik 1985) that it probably functioned as a sun shrine and not as a sun watcher's station or a "calendrical" site.

Furthermore, there is no compelling evidence for intentional lunar alignments at the three-slab site. Any potential lunar standstill moonrise alignments are serendipitous and were most likely unknown to the Chacoan petroglyph carvers, who were primarily taking advantage of the midday sun dagger phenomena to mark culturally meaningful solar events.

This brings up the question of who carved the two spiral petroglyphs at the site. Sofaer et al. (1979b:289-290) have suggested that "several factors show that the Anasazi inhabitants of Chaco developed the construct between AD 900 and 1300 (the approximate date of Pueblo abandonment of the canyon) and indicate that the specific time was between AD 950 and 1150, the period of greatest population and development in the canyon." This suggestion is not unreasonable, but neither is it supported by any archaeological evidence save the general stylistic analysis of the petroglyphs.

In fact, the only known archaeological remains found on Fajada Butte are two sites attributed to the late occupation of Chaco Canyon by Mesa Verde people--or Chacoans influenced by Mesa Verdean cultural traits--around AD 1220-1300 (Zeilik 1985:S72-73).

> This period is characterized by occupational use of cave sites and the tops of buttes (in addition to more traditional locations), a change in burial patterns, and the appearance of Mesa Verde Black-on-white and St. Johns Polychrome ceramics. Although once interpreted as evidence of a migration of Mesa Verde Anasazi into the Chacoan area, later analyses . . . suggest a change to increased economic interaction with Mesa Verde Anasazi rather than a migration. By 1300, however, Chaco Canyon was abandoned [Cordell 1984:102].

Though no artifacts have been found at the three-slab site itself, it was originally included as part of one of the two "Mesa Verdean talus-cliff sites" by the National Park Service Chaco Center Inventory surveyors in the summer of 1972 (Zeilik 1985:S72). This proximity does not prove that the three-slab site petroglyphs are from the Mesa Verde phase, but this must be taken as one of the best-supported hypotheses.

We may never know who carved the two spiral petroglyphs behind the sandstone slabs resting against the cliff face atop Fajada Butte. When and why they were created will likely remain the subject of controversy for years to come. What is certain, however, is that, with the mass of speculation and misinformation that has been purveyed to the public, the sun dagger will become a pseudoscientific legend of our time and times to come. Feeding the legend is the persistent confusion of the well-known Pueblo interest in moon watching involving the 29.5-day lunation cycle with the question of observing the lunar standstills and their 18.6-year period. As in the case of England's Stonehenge, we are witnessing an anthropological odyssey, the creation of a modern myth. This process is worthy of anthropological study in its own right. As romance novelist and heroine Joan Wilder discovered in a popular, tongue-in-cheek adventure movie of 1984, there is boundless excitement and adventure to be had in "Romancing the Stone."

REFERENCES CITED

Carlson, John B.

1983 The Selling of Fajada Butte: An Anacalypsis. Archaeoastronomy 6:156-160.

1984 Film review of "The Sun Dagger" (Produced by Anna Sofaer, Directed by Albert Ihde). Archaeology 37(2):78-79.

Cordell, Linda S.

1984 Prehistory of the Southwest. Academic Press, New York.

Frazier, Kendrick
1978 Solstice-Watchers of Chaco. Science News 114:148-151.
1979 The Anasazi Sun Dagger. Science 80 1:56-67.

Hadingham, Evan
1984 Film review of "The Sun Dagger" (Produced by Anna Sofaer, Directed by Albert Ihde). American Anthropologist 86:232-233.

Marshall, Leslie
1982 Secret of the Sun Dagger. The Washington Post Magazine, 18 July: 12-17.

Newman, Evelyn B., Robert K. Mark, and R. Gwinn Vivian
1982 Anasazi Solar Marker: The Use of a Natural Rockfall. Science 217: 1036-1038.

Reyman, Jonathan E.
1980 Letters: An Anasazi Solar Marker? Science 209:858, 860.

Sofaer, Anna, and Albert Ihde
1982 "The Sun Dagger." The Solstice Project. Bullfrog Films, distributor.

Sofaer, A., R. M. Sinclair, and L. E. Doggett
1982 Lunar Markings on Fajada Butte, Chaco Canyon, New Mexico. In Archaeoastronomy in the New World, edited by A. F. Aveni, pp. 169-181. Cambridge University Press, Cambridge and New York.

Sofaer, Anna, Volker Zinser, and Rolf Sinclair
1979a A Unique Solar Marking Construct of the Ancient Pueblo Indians. In American Indian Rock Art, Vol. V, edited by Frank G. Bock, Ken Hedges, Georgia Lee, and Helen Michaelis, pp. 115-125. Papers Presented at the Fifth Annual A.R.A.R.A. Symposium, The Dalles, Oregon, May 27-29, 1978. American Rock Art Research Association, El Toro, California.
1979b A Unique Solar Marking Construct. Science 206:283-291.

Williamson, Ray A.
1979 Field Report: Hovenweep National Monument. Archaeoastronomy 2(3): 11-12.

Williamson, Ray A., and Mary Jane Young
1979 An Equinox Sun Petroglyph Panel at Hovenweep National Monument. In American Indian Rock Art, Vol. V, edited by Frank G. Bock, Ken Hedges, Georgia Lee, and Helen Michaelis, pp. 70-80. Papers Presented at the Fifth Annual A.R.A.R.A. Symposium, The Dalles, Oregon, May 27-29, 1978. American Rock Art Research Association, El Toro, California.

Zeilik, Michael
1985 A Reassessment of the Fajada Butte Solar Marker. In Archaeoastronomy Supplement No. 9. Journal for the History of Astronomy 16:S69-S85.

Chapter 6

The Sun Personified:

Some Preliminary Observations Regarding Celestial Cosmology and the Ordering of Pueblo Indian Society

Theodore S. Jojola

The study of Pueblo Indian culture is not new. Many anthropologists, both archaeologists and ethnologists, have undertaken investigations and offered observations concerning the history, social structure, and behavior of the Pueblos (e.g., Dozier 1970; Tedlock 1979). The current study will not attempt to add new evidence to the literature, but some corrections of past misinterpretations will be offered.

The motivation for this paper came from my earlier attempt to reconstruct the history of outside contact for my home village, the Pueblo of Isleta (Jojola 1982). As I gathered information for that study, it became apparent that major historical incidents were highly vulnerable to subjective interpretation. I was impressed, for instance, by the sheer volume of facts collected, while at the same time I was surprised by the lack of a clear, persuasive argument in the way in which facts were presented. Except for the fixed chronological ordering of events, the information was subject to the widest and sometimes, I might add, the wildest range of interpretations. It was this variety that made the interpretations so suspect.

Hermeneutics, the science of interpretation or the skill of uncovering the meaning behind an author's phraseology, fits quite well into the research described here. Especially with regard to Pueblo Indian communities, which have been the subject of a plethora of studies, it is apparent that many of the conclusions of prior research should be reappraised. I rather fancy, for example, that Ruth Benedict's widely acclaimed treatise on Pueblo Indian Apollinarianism (Benedict 1934:79) was a product of her own romantic brand of humanism (Mead 1974:22-23). She viewed Zuni culture through the filter of her own philosophy and ascribed to the culture norms and behaviors that were based largely on her own idealism.

Reconstructed history is partly based on fact and partly built from the author's own interpretations. It is clear, therefore, that researchers are also "building their own story" in their attempts to describe their subjects' stories. Ruth Benedict was an anthropologist, but she was also a storyteller.

I felt it was important to examine this issue because the anthropological research entailed assumptions about my culture and my community. Since I was born and raised in Isleta, and therefore have been attuned to the cultural dimensions of Isletan society, the variation in anthropological interpretations of that society seemed to justify a reexamination of the facts on which those interpretations were based. In the end, my research resulted in a reconstruction and reinterpretation of facts that were quite shopworn.

I have raised this issue because it forms the basis of my reexamination of the Isleta Pueblo storytelling tradition. The facts regarding Isletan storytelling currently available in the anthropological literature are more than sufficient to enable new interpretations to be formulated. Rather than presenting arguments about which authors have developed the best theories, I will examine how they have developed their theories. An understanding of the derivation of theories is especially important when attempts are made to bridge the gap between a quantitative science, such as astronomy, and the qualitative realm of anthropology, and when these "bridges" (theories) in turn are used to construct a picture of how that society may have functioned prehistorically, as evidenced by archaeological remains.

THE STORYTELLING SETTING

A week or so before Christmas in 1884 Charles Fletcher Lummis arrived in a Pueblo village. In his own words,

> Three hours' walk thence to the south along the river--which was fairly alive with wild geese and ducks--brought us to the quaint Pueblo Indian town of Isleta. There was little dream in me, as we rambled through the strange little city of adobe, and interviewed its swarthy people, that this was some time to be my home [Lummis 1892b:140].

In 1888, a stroke of paralysis caused by overwork as city editor of the Los Angeles Times turned his attention, once again, to Isleta, and Lummis moved into the village in an effort to recoup his failing health. (Gordon 1972:7). The move represented a unique episode in the history of Isleta. It marked one of the few instances among all of the Pueblos in which an outsider was openly tolerated. During the four years that Lummis resided in the village he became enmeshed in village life and wrote extensively about its folk traditions (Lummis 1891, 1892a, 1892b, 1893, 1894, 1897a, 1897b, 1910, 1925).

Lummis was not alone in his efforts to study and describe Isletan life. Adolf Bandelier (1890), Elsie Clews Parsons (1920, 1921, 1928, 1932, 1939), Esther Goldfrank (1926, 1962, 1967), Volney H. Jones (1931), Janet B. Montgomery-McGovern (1932), Marie Hamilton Brown (1933, 1934), David H. French (1948), and more recently, Estellie M. Smith (1967, 1968; Smith and Leap 1969), William L. Leap (1970), and Florence Hawley Ellis (1979) have added to his observations. Lummis was the only one, however, to learn the language and interact freely with the people of Isleta; the other ethnologists used paid informants and generally lived in Albuquerque. The following excerpt is typical:

> Never had I encountered anything to compare with the hostility I met with in Isleta. After 9 frustrating days, in which I was given information so blatantly incorrect that I ignored it completely, my aloof

> Indian "family" introduced me to one of their close friends, "a good storyteller," they said [Goldfrank 1967:4].

This mode of research added an interesting twist to the accumulation of facts on Isleta. For the most part, the number of informants was small. In fact, Parsons, Goldfrank, Smith, and Leap corroborated the research of their predecessors, which is not surprising given that they shared many of these same informants. The identities of the informants were zealously protected from all but other researchers (Goldfrank 1926:70; Leap 1970:vii).

I do not intend to question the quality of their information. Indeed, if community controversy is any measure, then two publications in particular (Lummis 1910 and Goldfrank 1962) were apparently considered to be too accurate at the time they were first discussed by the Isleta populace. Nonetheless, the information did come from a small pool of native informants. It is clear that the overall value of the information lies not only in its accuracy but because it derives from distinctive viewpoints within the community.

ON STORYTELLING

> Over there to the East [is] the Sun
> Where [the Sun] comes out
> [is the] Father
> [is a] Man
> [The Sun] sees everything
> You [Sun] see [everything] clearly.
>
> Fillmore 1896:277

Traditionally, folktales are told during the winter. This season is more or less confined to the period from the fall equinox to the spring equinox (Parsons 1932:354). Several types of stories are told. During certain rites, the stories are oral histories. These do not vary and are passed down to successive practitioners in an exacting form and in a ritual cadence. The second type is what can be classified as "camp stories." These are laudatory and are built on individual deeds or personal recollections. Third are the folktales. As will be seen, their plots are built on perceived natural orders and around specific personae. In this paper I will limit my discussion to this folktale form.

Ethnolinguists have stressed the narrative style and structure of folktales as the basis for the transfer of cultural values and have downplayed the use of personification. In brief, they suggest that language is bounded by strict systematic relationships so that elements within a sentence can be described by the sound, the position (syntax), and the definition of the word (Leap 1970:2-3). Indeed, the oral history tradition of the Pueblos has been described as rather poetic in style. The success of the presentation is dependent on the skill of the narrator. Part of this skill, however, is the storyteller's ability to interweave his point of view within the story (Tedlock 1979).

Similarly, it has been implied that Pueblo stories are narrow treatments of ancient tales, with the diversity of stories being the result of variations on an ancient theme (Parmentier 1979:609). Once again, the variation in a story is attributed to the skills of the narrator, particularly to his ability to interweave a new variant within the original story based on his own personal motives, values, and experiences (Goldfrank 1926:78).

The above analyses might lead one to conclude that storytelling is a skill limited by a traditional structure and therefore is not a creative process. This may in part be true, especially with regard to oral histories, but one aspect that appears to have been overlooked is the character development that has taken place in the main personae.

Character development can best be exemplified by the personification of celestial phenomena and the use of these characterizations for creating new stories. A cursory summary of such characteristics for the Sun and Moon should serve to illustrate this point:

PERSONA		CHARACTER
Sun	Primary	Father
		Provider
		--domicile in east (Parsons 1939:92)
		--daily journey across sky with noontime as a still period used to visit his children (Parsons 1932:368)
		Protector
		--introduced to newborn (Parsons 1939:28)
		--kind to children (Goldfrank 1926:74)
		--does not scold children (Parsons 1921:155)
		--daytime guardian of the people (Lummis 1894:71)
		--source of longevity (Goldfrank 1962:88)
	Secondary	Man (Goldfrank 1926:74)
		--having bad/evil heart (Parsons 1932:340)
		--having two human sons, impregnating human mother with sun's rays, recognizing sons only through tests, having sons as companions (double rainbow, two red stars; Parsons 1932:342, 394-402)
		Younger brother of Dios (Parsons 1932:412)
Moon	Primary	First Woman
		--beautiful
		--pure heart
		Sun's first companion
		Sun's first wife
		Mother
		--nighttime guardian of the people (Lummis 1894:71)
		--blind in one eye, symbol of mother's dedication to family (Lummis 1894:72)
	Secondary	Beautiful Village Maiden
		--vain
		Best village housekeeper (Lummis 1894:57-64)

It should be noted that each character has primary and secondary personae. This multiplicity can be heard as a story evolves. Like humans, the personae mature both bodily and socially. In the mind of the listener, the supernatural become mortal. Indeed, the only trait that distinguishes the ordinary mortal from the supernatural being is the superhuman rate at which the latter matures (Parmentier 1979:609).

In the day-to-day context of growing up in Isleta, children soon become familiar with the traits of these celestial beings. Quite naturally, over the years the listeners develop a specific image of these personae. In a child's imagination they become real in the same way that a supernatural hero like Superman becomes real to Anglo children. Also in the same way, stories are built by creating new adventures and exploits.

In the course of identifying with these figures, the children are gradually being conditioned to assume their roles in the village. The long-term result of this emulation of celestial personae is the maintenance of the community's social structure. Individuals are expected to assume the persona of the deities during certain ceremonial occasions as well as in their day-to-day behavior.

Based on this perspective, it is clear that ethnolinguists may have developed a method of interpretation that is inflexible. Ethnolinguists have concluded that native people are only capable of memorizing basic plots and that these plots remain fixed throughout the generations. The use of personification devices in a story allows for much more creative liberty. In other words, the plot of the story is secondary to how the storyteller maintains the integral traits of the personae. These traits predominate in whatever situation the storyteller wishes to create.

Another aspect of storytelling is the attempt to balance the supernatural with human reality. The keepers of astronomical knowledge are important intermediaries between the two dimensions. Through their measurements they posit an image of natural order. They help to translate the physical aspects of the world into metaphysical ones. Through empirical processes, they begin by studying celestial cycles and movements. Although their methods may not be "mechanically" precise, they use three points of reference to cross-check the cyclical progress of a year: the passing of the solstices (Parsons 1921:155), the phases of the moon (Tyler 1964:161), and the seasonal positioning of recognizable constellations (Parsons 1932:288).

The relations apparent among the various celestial personae tend to reinforce the perception of the cyclical nature of celestial phenomena (Ortiz 1972:144). When important cycles were disrupted through human actions, disaster occurred in the form of drought and famine. In such cases the intervention of the Sun's two sons was often required to reestablish the cyclical rhythm. The overall effect of such cautionary stories on the listeners is a reinforcement of their sense of responsibility for the maintenance of the natural order. This differs markedly from the concept that native ceremonies were performed in order to invoke natural elements like rain as a response from the supernatural deities (Tyler 1964: xii).

Such ordering was also manifested in village planning. Fires were maintained in hearths located in the eastern portion of society houses (Gatschet 1891). Special symbolic elements were integrated into the buildings, and openings were positioned in the walls so as to measure the progress of the sun and constellations (Parsons 1939:92). Streets and houses were almost always oriented in the cardinal directions (Ellis 1975:85). It has even been implied that pueblo buildings were limited to five stories so as not to exceed the five levels mentioned in the emergence stories (Harrington 1930:79).

This relationship between humankind and the supernatural can be summarized in a concise cosmological, four-dimensional model of the world. On a horizontal plane are the cardinal directions. This dimension represents the field of human activity. The vertical plane of upper and lower levels is the second dimension, which serves to fuse the past, present, and future. At the center is a special point that is an inward direction. It is a negative infinity that represents an accessway allowing objects to transcend the third and fourth (i.e., supernatural) dimensions.

CONCLUSION

In conclusion, archaeoastronomers, archaeologists, cultural anthropologists, and ethnologists would do well to reexamine their own motivations as storytellers. In certain regards it is safe to say that important evidence has gone unrecognized because of the perspectives social scientists carry into the field and into their writings. Correspondingly, the effective use of oral storytelling hinges on the ability of Pueblo keepers of astronomical knowledge to translate their observations using a frame of reference that is familiar to the community. Thus, the distinction between the researcher and the Pueblo astronomer is not so great. Both must explain facts within their own particular frame of reference.

On a personal note, I am reminded of an incident that will help to illustrate this point. A Navajo friend of mine and I were having coffee and discussing her first trip to Chaco Canyon. She told me, "You'll never believe what I heard." Apparently a park ranger was interpreting the sites for a tour group, and he concluded that "the Chaco Canyon civilization had to be a slave society. How else can anyone explain the monumental scale of what we see?" I would have liked to have asked him if he thought that Albuquerque was built in a day.

We are all intrigued by "dead" civilizations. They are easy to explain because they are dead (and can't talk back). To avoid this trap, we must assess our own points of view in light of the stories we construct. Otherwise, I am afraid that there will be little difference between our stories and that of the park ranger. Like the old Spiderwoman of Pueblo folklore, he was spinning his own web--and catching unwary tourists.

REFERENCES CITED

Bandelier, Adolf F.
1890 _Final Report of Investigations Among the Indians of the Southwestern United States, Carried on Mainly in the Years from 1880 to 1885_, Part 1. Papers of the Archaeological Institute of America, American Series, Vol. 3. John Wilson and Son, Cambridge, Massachusetts.

Benedict, Ruth F.
1934 _Patterns of Culture_. Houghton-Mifflin, New York.

Brown, Marie Hamilton
1933 Tales of Isleta: Sketches of Revealing Characteristics of Isleta. _New Mexico Quarterly_ 3:9-17.
1934 Tales of Isleta: Sketches of Revealing Characteristics of Isleta. _New Mexico Quarterly_ 5:281-290.

Dozier, Edward P.
1970 Making Inferences from the Present to the Past. In _Reconstructing Prehistoric Pueblo Societies_, edited by William A. Longacre, pp. 202-214. University of New Mexico Press, Albuquerque.

Ellis, Florence Hawley
1975 A Thousand Years of the Pueblo Sun-Moon-Star Calendar. In Archaeoastronomy in Pre-Columbian America, edited by Anthony F. Aveni, pp. 59-88. University of Texas Press, Austin.
1979 Isleta Pueblo. In Southwest, edited by Alfonso Ortiz, pp. 351-365. Handbook of North American Indians Vol. 9. Smithsonian Institution, Washington, D.C.

Fillmore, John Comfort
1896 Two Tigua Folk Songs: The Song of the Sun and the Song of the Wheel Dance. In Land of Sunshine 4(6):273-280.

French, David H.
1948 Factionalism in Isleta Pueblo. In Ceremonial Patterns of the Greater Southwest, edited by Ruth M. Underhill, pp. 1-48. Reprinted as Monograph of the American Ethnological Society. J. J. Augustin, New York.

Gatschet, Albert S.
1891 A Mythic Tale of the Isleta Indians. American Philosophical Society Proceedings 29:217-219.

Goldfrank, Esther Schiff
1926 Isleta Variants: A Study in Flexibility. Journal of American Folklore 34:70-78.
1962 Isleta Paintings. Bureau of American Ethnology Bulletin 181. Smithsonian Institution, Washington, D.C.
1967 The Artist of Isleta Paintings in Pueblo Society. Smithsonian Contributions to Anthropology 5. Washington, D.C.

Gordon, Dudley
1972 Charles F. Lummis: Crusader in Corduroy. Cultural Assets Press, Historical Society of Southern California.

Harrington, John P.
1930 Buildings of Native Americans. PanAmerican Magazine 43(2):79-90. Geographic and Historical Society of the Americas, Washington, D. C.

Jojola, Theodore S.
1982 Tribal Survival and Outside Contact: The Spanish Colonial and American Territorial Occupation of the Bontoc and Isleta Villages. Unpublished Ph.D. dissertation, Department of Political Science, University of Hawaii, Manoa.

Jones, Volney H.
1931 The Ethnobotany of the Isleta Indians. Unpublished Master's thesis, University of New Mexico, Albuquerque.

Leap, William L.
1970 The Language of Isleta, New Mexico. Unpublished Ph.D. dissertation, Department of Anthropology, Southern Methodist University, Dallas.

Lummis, Charles Fletcher
1891 A New Mexico David and Other Stories and Sketches of the Southwest. Scribner's, New York.
1892a Some Strange Corners of Our Country. Century, New York.
1892b A Tramp Across the Country. Scribner's, New York.
1893 The Land of Poco Tiempo. Scribner's, New York.
1894 The Man Who Married the Moon and Other Pueblo Indian Folk Stories. Century, New York.
1897a The Enchanted Burro: Stories of New Mexico and South America. Way and Williams, Chicago.
1897b The King of the Broncos and Other Stories of New Mexico. Scribner's, New York.
1910 Pueblo Indian Folk Stories. Century, New York.
1925 Mesa, Cañon, and Pueblo: Our Wonderland of the Southwest. Century, New York.

Mead, Margaret
1974 Ruth Benedict. Columbia University Press, New York.

Montgomery-McGovern, Janet B.
1932 A General Survey of Isleta Pueblo with Special Reference to Acute Transitional Conditions. Unpublished Master's thesis, University of New Mexico, Albuquerque.

Ortiz, Alfonso, editor
1972 New Perspectives on the Pueblos. University of New Mexico Press, Albuquerque.

Parmentier, Richard J.
1979 The Pueblo Mythological Triangle: Poseyemu, Montezuma, and Jesus in the Pueblos. In Southwest, edited by Alfonso Ortiz, pp. 609-622. Handbook of North American Indians Vol. 9. Smithsonian Institution, Washington, D.C.

Parsons, Elsie Clews
1920 Notes on Isleta, Santa Ana, and Acoma. American Anthropologist 22:56-69.
1921 Further Notes on Isleta. American Anthropologist 23:149-169.
1928 The Laguna Migration to Isleta. American Anthropologist 30:602-613.
1932 Isleta, New Mexico. In Forty-Seventh Annual Report of the Bureau of American Ethnology, pp. 201-446. Smithsonian Institution, Washington, D.C.
1939 Pueblo Indian Religion, 2 vols. University of Chicago Press, Chicago.

Smith, M. Estellie
1967 Power and Politics: More Factionalism at Isleta Pueblo. Paper presented at the Sixty-Sixth Annual Meeting of the American Anthropological Association.
1968 Governing Systems and Culture Change Among the Southern Tiwa. Paper presented at the Sixty-Seventh Annual Meeting of the American Anthropological Association.

Smith, M. Estellie, and William L. Leap
1969 The Politician at Isleta Pueblo: An Ethnolinguistic View. Paper presented at the Sixty-Eighth Annual Meeting of the American Anthropological Association.

Tedlock, Dennis
1979 Zuni Religion and World View. In Southwest, edited by Alfonso Ortiz, pp. 499-508. Handbook of North American Indians Vol. 9. Smithsonian Institution, Washington, D.C.

Tyler, Hamilton A.
1964 Pueblo Gods and Myths. University of Oklahoma Press, Norman.

Chapter 7

Light and Shadow, Ritual, and Astronomy in Anasazi Structures

Ray A. Williamson

THE MAIDEN AND THE SUN[1]

A sudden flash.
The shaft of light,
Entering the darkened room,
Struck deep within,
Engendering new life in one
Who moved with delight.

Twins.

The old man was mad
When he learned.
He had safely locked away
In the fourth inner room
The Maiden,
His daughter.

But now this.

She, for her part,
Grew round.
But she never forgot
Him,
Sun Father,
And his awesome power.

She dreamed.

Later, long after she
Had left them lying,
The Twins,
And turned away,
Even in her shame
She pleasured.

Two boys.

INTRODUCTION

Historical Pueblo tales, mythology, and direct ritual practice reveal ample evidence of the importance of light and shadow to the Pueblo people. Observations of the interplay of light and shadow were part of Pueblo behavior at calendrically important times of the year, whether they involved direct sunlight, as in the one-time Zuni practice of keeping a calendar by watching sunlight fall through a porthole on a predetermined mark on the opposite wall (Cushing 1967:40), or indirect, as in the Isleta ceremony of "pulling down the sun" by use of a crystal (Parsons 1930:292-293). These practices, and the stories of the role of the sun in impregnating a Pueblo maiden by entering through a porthole or crack in the wall of a room, may afford us clues by which to interpret Anasazi structures that demonstrate evidence of purposeful astronomical alignment.[2]

In this paper I will present and interpret three diverse examples of Anasazi structures, each of which appears to have been intentionally aligned with astronomical phenomena. Each illustrates what I believe to be a different facet of Anasazi astronomy; each also exhibits features related to historical Pueblo practice. Part of my interpretation depends directly on the evidence afforded by the historical Pueblos. Part requires reasoning directly from the forms and orientations of the structures themselves as integral components of their environment.

CASA RINCONADA, CHACO CANYON

Casa Rinconada is a large, complex circular structure about 20 m across and 3 m deep. Set apart from the other buildings of similar construction in the canyon and located on the south side of the wash, it was apparently built in the late eleventh century during the height of Chaco Canyon development and social organization. Earlier papers (Williamson 1982; Williamson et al. 1977) present the details of the astronomical alignments observed in this kiva (Figure 7.1). The two primary types of astronomical alignments are cardinal orientations of the major and minor axes, as well as of the symmetrically placed internal features, and a summer solstice sunrise line between the northeast wall opening and a lower niche along the northwest side of the interior wall (Figure 7.2).

No one doubts that Casa Rinconada was a religious structure of considerable importance. Just how it was used, and precisely what place it assumed in the religious lives of the people of Chaco Canyon, is not known. Nor do we have any direct clues from the historical Pueblos. Yet investigations of the nature of its putative alignments may help us to understand the role of astronomy in the ritual life of the Chaco Canyon Anasazi. Indeed, elsewhere (Williamson 1982, 1984) I suggest that the particulars of the construction of Casa Rinconada reflect elements of historical Pueblo ritual.

Cardinal Orientations

After making a detailed map of Casa Rinconada's interior by means of a transit and steel tape, my colleagues and I concluded that its builders had paid considerable attention to the form and structure of Casa Rinconada and had taken unusual pains to be precise.[3] I will summarize these features here and suggest their relationship to Pueblo ritual notions:

1. The kiva as a whole is a carefully made circle. The center, defined by taking midpoints between pairs of niches across the kiva, falls within a circle less than 10 cm in diameter.[4]

2. The angular distance between each of 28 small wall niches and its next neighbor is highly regular. The center-to-center average angular distance is 11 degrees ± 19 minutes. The probable error of ± 19 minutes corresponds to only 5 cm, or 3 percent of the distance between niches.

3. The four holes that served as post sockets for the logs that originally supported the massive roof nearly form a square, the center of which falls within the 10 cm circle defining the center of the circular room.

4. The sides of this square are nearly cardinal lines.

5. The major and minor axes of symmetry are nearly cardinal lines. These are defined respectively as the center line between the north and south doorways and the line between small niches 8 and 22, counting clockwise from the north.

6. The floor vaults on either side of the major axis of symmetry are oriented along north-south lines.

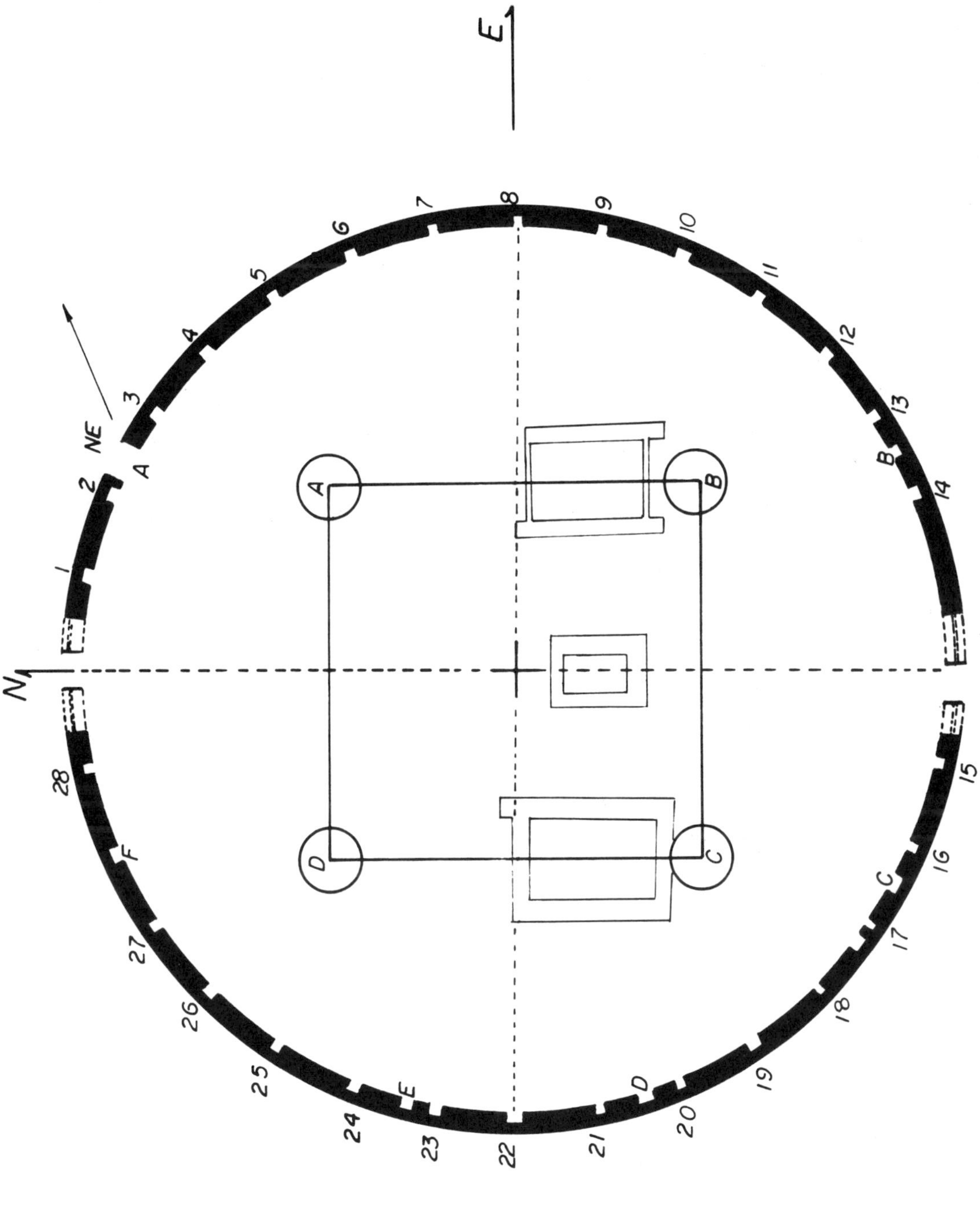

Figure 7.1. Simplified ground plan of Casa Rinconada

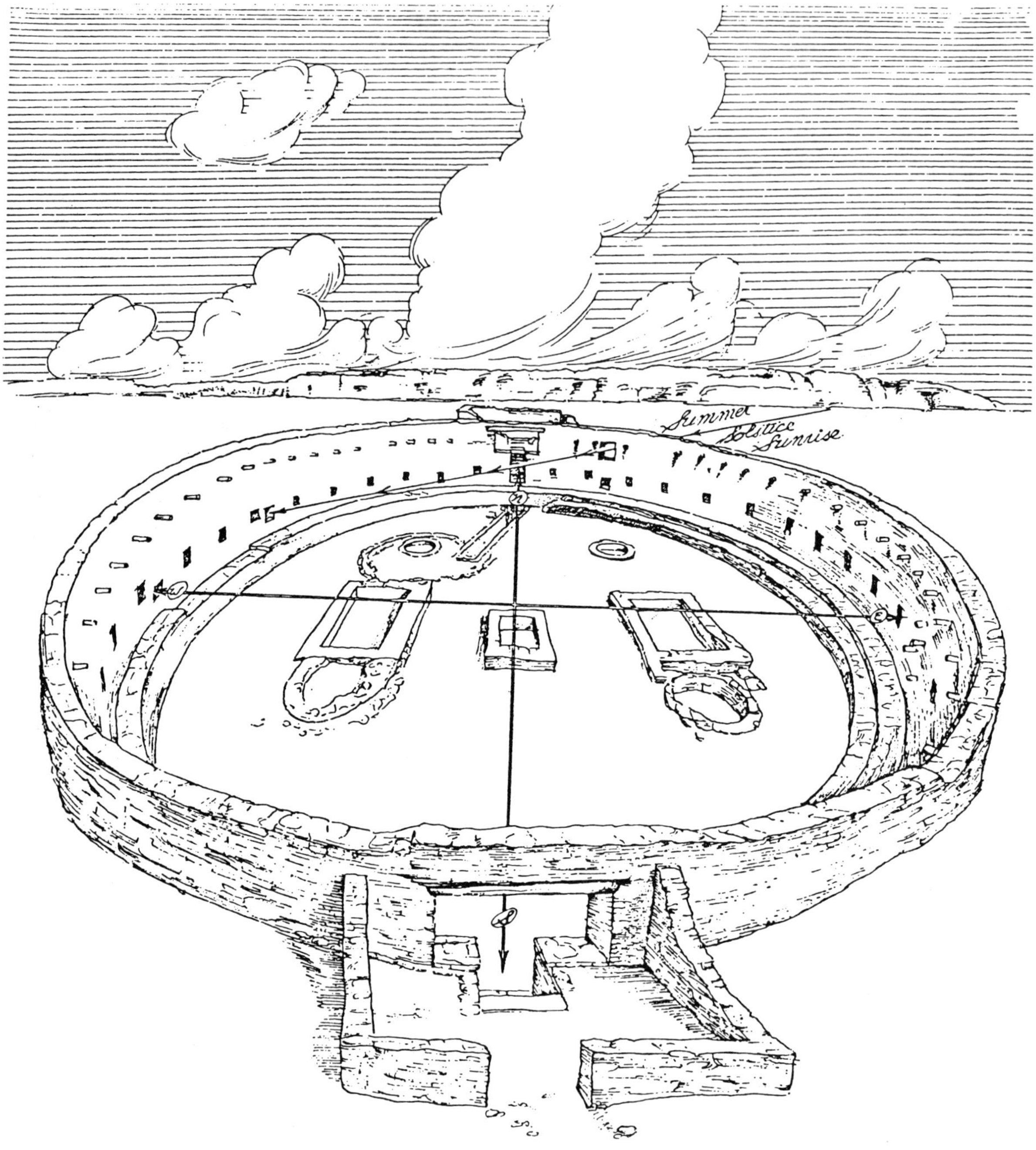

Figure 7.2. View of Casa Rinconada from the south. Illustration by Snowden Hodges

7. Lines joining the large lower niches A, B, C, and F form a rectangle, the sides of which are nearly cardinal lines.

8. The kiva as a whole is located due west of the edge of a mesa. At sunrise near the equinox, the sun appears momentarily to the north of the mesa edge and casts a shadow across the south half of the kiva before disappearing again behind the mesa as the sun rises to the south.

I suggest that these attributes indicate the builders' intention to place the kiva in the "proper" orientation with the cosmos, and to make it an image of time and space consistent with Anasazi notions of these two aspects of being (Williamson 1982). Support for this view is found in several aspects of Pueblo ritual, but perhaps most clearly in the description of the first kiva from the emergence myths of the Acoma, a Pueblo group that some (e.g., Ellis 1967) have argued are descended in part from the Anasazi of Chaco Canyon.

> When they built the kiva, they first put up beams of four different trees. These were the trees that were planted in the underworld for the people to climb up on. In the north, under the foundation they placed yellow turquoise. Prayer sticks are placed at each place so the foundation will be strong and will never give way. The walls represent the sky, the beams of the roof (made of wood of the first four trees) represent the Milky Way. The sky looks like a circle, hence the round shape of the kiva. . . .
>
> On the north is a hollow dug-out place that represents the door of North Mountain, East Mountain, West Mountain, Sun and Moon [Stirling 1942:19].

Conceptually, the first kiva at Acoma sounds very like Casa Rinconada, even to the detail of having a "hollow dug-out place" in the north (a long trench that leads to a staircase under the north entrance). Even if the kiva of Acoma myth has no direct relationship to Casa Rinconada, the myth at least gives us a starting place for thinking about Casa Rinconada and what its builders might have had in mind.

Other Pueblo historical data are also of interest. When the Hopi priest prays, he casts sacred cornmeal toward the four cardinal directions in an antisunwise motion. As Ortiz (1972:143) has pointed out in another context, although the priest's gesture is outward, it is in supplication to Sun Father, whose power emanates from the rim of the world toward the center. His gestures convey the spatial image of a square whose diagonals intersect at the center where the priest stands.

Table 7.1 summarizes the features evident in Casa Rinconada and the historical Pueblo conceptions to which they may be related. Some of these features are peculiar to Casa Rinconada, while others are evident in most kivas. Of particular interest to my argument, perhaps, is the number and placement of the four sockets for the roof posts. The presence of four sockets reflects the Pueblo ritual interest in the number four, which can represent either the cardinal (perpendicular) directions or the solstice directions. Structurally, it might have been much easier to support the roof with six or eight posts. But here, as in the great kivas of Pueblo Bonito and Chetro Ketl, the number of posts is four. As the Acoma emergence myth makes clear, the use

Table 7.1. Kiva features possibly related to Pueblo world view

Elements of Pueblo World View	Related Kiva Features
Directionalism (NW, SW, SE, NE)	Four posts
Importance of center -- space and time	Kiva placement Kiva orientation Centers of features
Twofold symmetry of space and time (dualities)	Kiva symmetry (time and space)
Underworld an image of this world	Overall kiva structure

of four beams to support the roof is explicitly related to the four directions, a conception that is equivalent to the Navajo use of posts in each of the four directions to support the hogan (Haile 1947:49) or the Pawnee use of four poles to support the earth lodge (Chamberlain 1982).

In Casa Rinconada, the symmetrical placement of other floor and wall features, such as the floor vaults and the wall niches, reinforces the symmetry of the four roof posts. Although the Navajo and Pawnee structures mentioned here are houses, whereas Casa Rinconada is a ritual structure, the analogy is still apt. In both the Navajo and Pawnee cases the forms of the structures are prescribed by origin myths (Dorsey n.d.; O'Bryan 1956:13), as is the form of the kiva in the Acoma myth. In addition, the kiva can in an important sense be thought of as a home, for it is explicitly understood as being symbolic of the original home from which the Pueblo people emerged and to which they return at death. When climbing the ladder (or, in the case of Casa Rinconada, presumably the staircase) out of the kiva, one symbolically reemerges from the underworld (Stirling 1942:20).

In some sense both the cardinal (perpendicular) and solstice directional sets are present in Casa Rinconada if one is willing to consider the corners of the square as symbolic of the solstice directions. For the historical Pueblos, it is the solstice directions that are critical to ritual practice rather than the cardinal orientations. The emphasis on cardinal orientations both at Casa Rinconada and in Pueblo Bonito across the canyon (Williamson 1978; Williamson et al. 1977) may reflect an aspect of Anasazi ritual practice that has since been lost.

The Sun's Presence

Light and shadow play an interesting role in Casa Rinconada. The northeast part of the kiva contains an opening through which sunlight enters the structure at about 30 minutes after sunrise on and around the summer solstice.[5] The existence of this alignment, along with the other astronomical alignments in the kiva, leads me to believe that the summer solstice alignment was intentional. Because so little of the original opening was left at excavation (Vivian and Reiter 1960), however, it is impossible to determine whether this light-admitting feature served a calendrical (anticipatory) or a celebratory function.[6] My personal opinion is that if it was intentionally placed in the kiva, the opening probably served to celebrate, rather than to announce, the solstice (note, however, Zeilik's counter arguments in Chapter 3).

It seems to me that the best way to understand this alignment is not in terms of calculating a calendar but in terms of what Eliade (1959:11) calls a hierophany, literally a "sacred showing." Today, even with the roof gone, the appearance of the sun through the port is an impressive sight. Consider the effect of sunlight streaming into the kiva when it had a roof. In other words, this alignment appears to be more closely related to the myths of the sun and the Pueblo maiden presented in the poems that open and close this chapter than to calendrical calculations. In Casa Rinconada, we might imagine that Sun Father demonstrated his power directly to the participants in the kiva at sunrise. The historical Pueblo groups recognize and use the obvious power of sunlight in these ways. In another example, the Acoma myth uses the symbol of light in the upper world, as opposed to the darkness in the lower world ("they were living in the dark without knowing each other by name" [Stirling 1942:1-2]), to symbolize the passage from ignorance to knowledge. It seems reasonable to suggest that the Anasazi did the same.[7]

The effects of light and shadow are part of the phenomena of the world, and as Ortiz (1972:143) again reminds us, for the Pueblo "everything in the cosmos is believed to be knowable and, being knowable, controllable." Further, I believe that the Anasazi builders of Casa Rinconada used the placement of the kiva and incorporated appropriate structural details to control the entrance of the summer solstice sunlight for maximum ritual effect.[8] Indeed, I am reminded here of the control of light and shadow for dramatic purposes apparent in the Isleta ceremony of "pulling down the sun" at the summer and winter solstices (Parsons 1930:292-293). At Zuni, a hole was placed in the roof of the ceremonial rooms of the Rain Chiefs so as to enable the light of the sun to fall on the altar (Parsons 1939:378). The Hopi similarly used the sun's light in the early days of this century (Curtis 1922:133; Parsons 1939:607). In this way they directly and symbolically joined the celestial powers with the earthly ones. Even today, the Hopi celebrate the arrival of the solstice when sunlight streams through a port in a Walpi kiva (Frazier 1978).

Space and Time in Casa Rinconada

If, as the Acoma myth suggests, the builders of Casa Rinconada considered the rim of the kiva as symbolic of the rim of the world, then the overall structure of Casa Rinconada, organized as it is about the cardinal directions, unifies time and space in a single symbolic structure. In noting the passage of the sun along the horizon during the year, the careful observer cannot avoid seeing and incorporating into his or her thinking the intimate relationship between time and space.[9] In the kiva as it now stands, one point in the cyclical march of time is fixed by the summer solstice alignment. The equinoxes can be considered to be marked directly by the shadow cast by the eastern mesa.[10]

Elsewhere my colleagues and I have speculated that there might have originally been a winter solstice sunset alignment that would have completed the structural inclusion of the major turning points of the calendar (Williamson et al. 1977:209-210). Because the kiva has been reconstructed, and because there is no indication of another large port in the excavation records, we cannot know whether such an alignment was originally included. Even if the winter solstice was not marked, however, the orientation of the kiva as a whole, plus the symmetric and regular placement of internal kiva features, reflects the division of time along the "rim of the world." Between the vernal equinox and the autumnal equinox, the sun always rises and sets north of the minor axis of symmetry. The opposite is true for the period between the autumnal equinox and the vernal equinox. The summer and winter duality, a division of major importance to historical Pueblos, is represented in the kiva's structure.[11]

CAJON GROUP, HOVENWEEP NATIONAL MONUMENT

I will now shift the focus of this discussion about 200 km northwest of Chaco Canyon and roughly a century later to discuss a group of buildings on the southern boundary of Cajon Mesa in southeastern Utah. The builders of these structures were culturally related to the Mesa Verde Anasazi and used similar building techniques--loaf-sized blocks of quarried sandstone laid with mud mortar. Both Hopi myth and recent archaeological investigations (Winter 1976) relate the Anasazi of Hovenweep to the Hopi. In contrast to the Mesa Verde Anasazi, these people built smaller dwellings and located them at the heads of canyons near a secure source of water rather than in cliff alcoves.

The principal buildings in the Cajon group are a block of rooms on the east rim of the canyon, a larger roomblock on the west rim, and a tower with two ground-floor rooms, also on the west side. Below the rim in the canyon are the remains of at least one tower as well as several rooms connected with the eastern roomblock. The buildings on which I wish to focus are those on the west side. While surveying the towers of Hovenweep in order to determine whether they contained astronomical alignments, my colleagues and I were amazed to discover that these two buildings interacted so as to create a solar calendar (Williamson 1984).

Immediately after the autumnal equinox, the shadow of the larger roomblock begins to cross the tower just before sunset. As fall progresses and the sun moves south, the shadow moves progressively northward across the tower each evening until, at the winter solstice, one of the eastern corners of the roomblock casts a shadow that falls in a straight line between the roomblock corner and the southwestern corner of the tower (Figures 7.3 and 7.4). After the winter solstice, the shadow moves south each day as the sun moves north until the vernal equinox. At equinox the shadow of the northeast corner of the roomblock falls parallel to the southern wall of the tower, which is oriented nearly east-west.

As the sun continues to move north along the horizon, the shadow of the tower falls across the roomblock at sunrise, but the roomblock no longer casts a shadow across the tower at sunset. The sunrise shadowcast continues throughout the spring and summer until the autumnal equinox, when the full cycle is complete and the shadowing occurs at sunrise rather than sunset. We do not know just how the summer solstice sunrise shadow might have fallen on the roomblock because the northern end of the tower is in ruins, and without excavation we cannot determine the original extent of the tower wall. Measurements taken in 1983 and 1984 during research expeditions with Crow Canyon students unfortunately left the question undecided (Williamson 1984; Williamson and Young 1983).

These alignments are highly accurate (i.e., they are defined to within 2.5 cm linear distance along the wall), and they occur on a scale that would make them visible to the entire pueblo. If the inhabitants of the Cajon group were interested in defining a precise calendar, painted markings on the outer walls

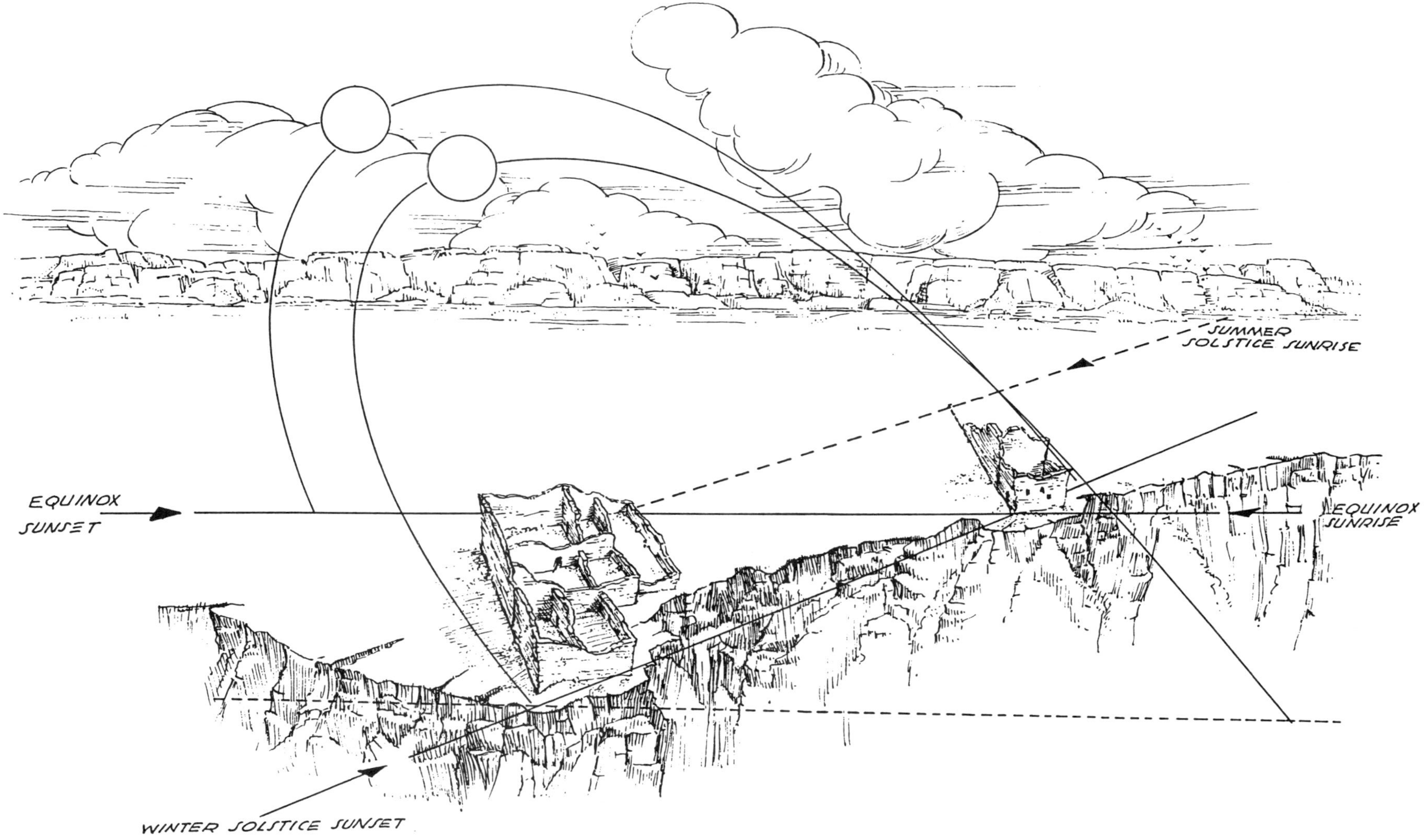

Figure 7.3. Major solar alignments indicated by shadows at Cajon Ruin. Illustration by Snowden Hodges

Figure 7.4. View of tower and western roomblock from the east, Cajon Mesa

of the tower and roomblock would have permitted them to follow the progress of the shadows across these structures as the sun moved through its yearly cycle.

Because the sun moves nearly a solar diameter along the horizon from day to day around the equinox, the shadow that is cast by the northeast corner of the roomblock, for example, would shift about 12 cm north from one day to the next along the tower wall. This would have made identification of the day of equinox quite easy. Because the sun moves very slowly at the solstice, on the other hand, it would not have been possible to use similar techniques to detect the precise day of winter solstice. Observers would have had to use marks at the appropriate positions along the wall several days in advance of the solstice, as Zeilik (Chapter 3) notes is done in the historical pueblos, or they would have had to use additional, independent observations.

The important point is that this shadow casting would have been a public display, knowledge of which was not necessarily limited to a small coterie of initiates. This, too, is consistent with historical Pueblo practice. Although certain individuals were charged with the duty of watching the sun to determine the calendar, calendrical information, especially that connected with planting and harvesting or with the timing of ceremonies, was fairly common knowledge. As Crow Wing, the Hopi, wrote in his journal of April 18, 1921, "We think the Sun-Watcher is not a very good man. He missed some places, he was wrong last year. All the people think that is why we had so much cold this winter and no snow" (Parsons 1925).

Ellis (1975) argues that the Puebloans tend to emphasize the usual rather than the unusual. Just as they deemphasize the role of the individual and emphasize the community, the historical groups pay great attention to the regularities of the universe. If my interpretation of the Cajon buildings is correct, it seems as if the builders of the Cajon structures made use of the commonplace occurrence of long shadows at sunrise and sunset to create a specific functional effect in delineating a yearly solar calendar.

Here again, as at Casa Rinconada, the builders brought together the dual aspects of time and space, although in a radically different form than in the great kiva. Only on the equinox are the shadows cast at sunrise (east to west) and sunset (west to east) the reverse of each other. At Cajon the builders apparently used that fact to switch from sunrise to sunset calendrical markers on the equinox. Because of this switch from eastern to western control of the shadows, it is interesting to speculate that ceremonial control of the pueblo might also have changed hands at the equinoxes, as it does today in some Rio Grande pueblos (Ortiz 1969).

HOVENWEEP NATIONAL MONUMENT

As my final example of the interaction of light and shadow in calendrically significant Anasazi structures, I will discuss a petroglyph panel located in Holly Canyon, about 20 km northwest of the Cajon group. This panel is the site of a striking and moving interplay of light and shadow after sunrise on the summer solstice.

The petroglyphs are pecked on the northern side of one of several large sandstone boulders about 100 m south of Holly House on the west edge of the canyon. The panel's principal elements (Figure 7.5) are two spirals, a Pueblo "sun symbol" of three concentric circles with a

Figure 7.5. Petroglyph panel, Holly Canyon, from the west.
Illustration by Snowden Hodges

central dot (Ellis and Hammack 1968), a wavy line that appears to represent a plumed serpent, and two circles, each with a central dot and connected by a curved line. An overhang above the petroglyphs protects them from the weather.

As Williamson and Young (1978) have previously pointed out, such symbols are directly related in Hopi myth and ritual with light and warmth, water, and increase of crops. The sun and the plumed serpent are also related to Hopi ceremonies at the winter solstice and the vernal equinox. We originally investigated the site because it appeared to be directly related to the local planting and harvesting cycle. The boulders form a natural corridor aligned roughly east-west. Although the western horizon is blocked by another large boulder, the eastern horizon is open to sunrise from ca. February 20 (around the proper time to begin preparing the fields) until roughly the vernal equinox. From then until the summer solstice the sun peeks through at intervals between the boulders, possibly allowing the site to be used to define a solar calendar of suitable times for planting. As the sun moves south, the same phenomena could be used to predict the harvest.

Several years after my initial examination of this panel I discovered that there was a significant alignment on and near the summer solstice (Williamson 1984:92-103). About 45 minutes after local sunrise a narrow sliver of sunlight begins to enter the long, narrow corridor through a natural slit formed by two boulders and begins to cut across the left-hand spiral. As the streak of light elongates and reaches across the second spiral, a second light beam begins as a tiny point of light far to the right of the sun symbol. It too extends, but it moves to the left toward the sun symbol and the first pencil of light. In a minute or so the second narrow streak of light cuts the sun symbol just below the center. Shortly afterward, the two extended streaks of light, which seem serpentlike in their movements, join each other between the sun symbol and the spirals, broaden, and move downward across the rest of the petroglyph panel. It takes only 7 minutes for this compelling drama of light and shadow to complete its course, but in that time even the modern observer is drawn into the drama of the event (Figure 7.6a-h).

This intricate and fascinating display of light is caused by the interaction of the moving sun and the complicated three-dimensional geometry of the boulders--boulders placed there by nature and used by human beings. The streaks of light were entirely natural phenomena until they were noticed by the Anasazi, who placed the glyphs strategically on the face of the southern boulder.

I suggest that these petroglyphs and their interactions with light from the rising sun served both ritual and calendrical purposes for the Anasazi who placed them there. The patterns of light and shadow that play along the panel, when watched carefully for a sufficient time, are capable of allowing one to devise a calendar suitable for planting and harvesting. In fact, I believe it would be possible, by watching the forms of the streaks of light several weeks prior to the solstice, to anticipate the arrival of the solstice with an accuracy of within two or three days. To test my hypothesis adequately will require further detailed observations, but my photographic records indicate that it should be possible to discriminate with this degree of accuracy.

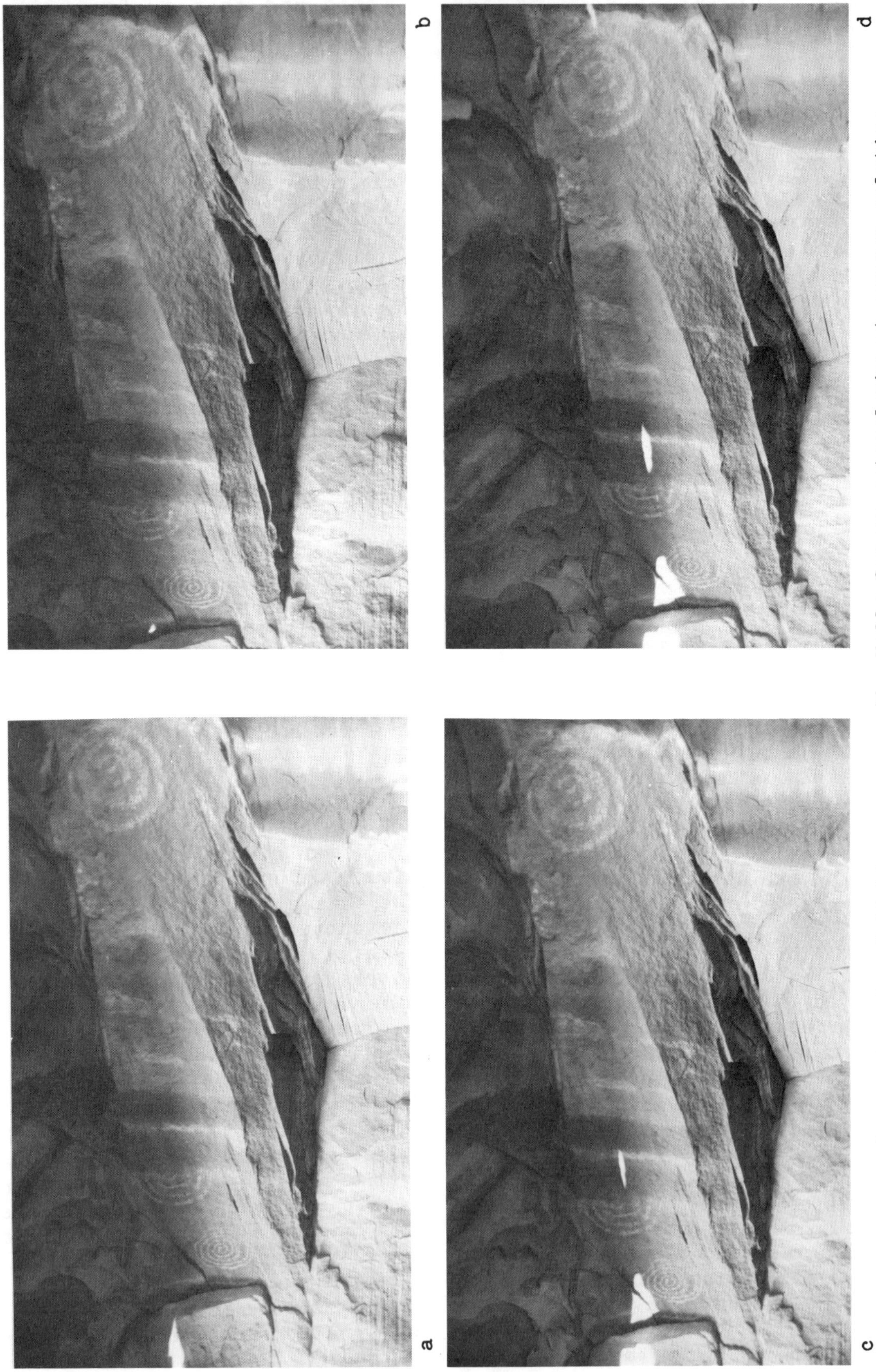

Figures 7.6a-d. Streaks of light as seen on the Holly Canyon petroglyphs at summer solstice, approximately 45 minutes after local sunrise

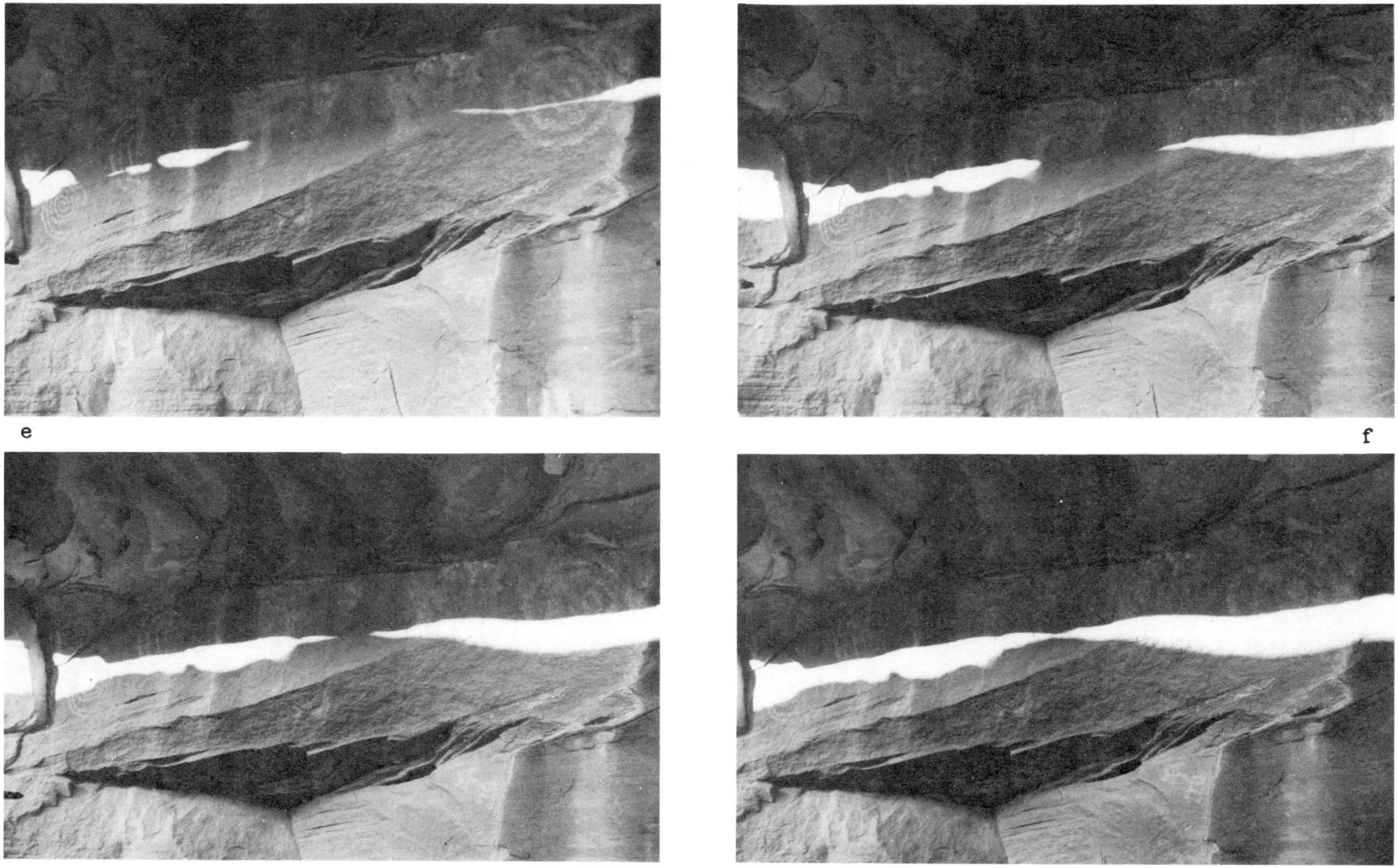

e f g h

Figures 7.6e-h. Streaks of light as seen on the Holly Canyon petroglyphs at summer solstice, approximately 45 minutes after local sunrise

In some respects the straightforward calendrical information the panel could impart is only a small part of the potential meaning the petroglyphs may have had for the artist or artists who placed them there. As Williamson and Young (1978) have emphasized, the images on the panel are similar to those used in Hopi ritual observance at the solstices and equinoxes. They also reflect the grand dualities of Pueblo culture--light and water, sun and serpent, and perhaps the Twins, sons of the Sun.

The Twins are extremely important to Hopi and Zuni mythology. At Zuni, at least, they represent the morning and evening stars. Because they are the result of the union of an earthly mother and a celestial father, they also embody another duality. In most of the myths (e.g., Benedict 1935:49), after growing for four days, or sometimes four years, the Twins set out to seek their father, the sun. Upon their arrival at his house, the Sun first tests them and then claims them as his own children after seeing how well they survive his arduous tests. Thus, in the myth we first have the Sun coming to earth to couple with an earthly maiden. The product of that union is also a duality. The Twins complete the cycle by going to their father's house in the sky to claim him as their parent.

The other duality illustrated in the panel is that between the sun and the plumed serpent. The Hopi make the tension between these two symbols of light and water explicit in their winter solstice ceremony, in which Fewkes (1893:272) suggests that "the evil influences of the great snake are met by the prayers to his effigy." Later, near the vernal equinox, the plumed serpent, Palulukong, appears and violently overthrows young corn plants that are set out before him. Only after being offered prayer meal does Palulukong withdraw, to interfere no longer with the growth of the young corn. After a long wait in the ceremony the opposition between the sun and water is dramatized directly when the Hopi Sky God stands near the center of the kiva and struggles with an effigy of Palulukong. The serpent realistically twists around the masked god and fights with him. Then, without warning, it sweeps down upon a group of young corn plants and knocks them over amid loud cries of protest from the audience. Palulukong wins a momentary victory before he is subdued by the Sky God.

CONCLUSIONS

Thus far I have tended to treat both the historical Pueblo people and their ancestors as if all Pueblos engaged in the same astronomical practices. There are considerable differences in practice among the historical groups, however, and we would expect the Anasazi to have had a multiplicity of ways of defining a calendar and celebrating calendrically important events as well.

In my view, we do not yet have sufficient data to differentiate with any precision among the different Anasazi groups in terms of their astronomical practices; perhaps we never will. But we can at least attempt to gather and organize the data. I am convinced by my previous studies (Williamson 1984) that the people of Hovenweep were using some of their structures to keep a calendar. It is not yet clear to me that the Chaco Canyon people were doing the same. With the possible exception of the site at Fajada Butte (Sofaer et al. 1980; Chapter 4, this volume), whose Anasazi origin is uncertain and whose alignments are of a radically different sort than the ones discussed in this paper, none of the alignments we have found so far in Chaco Canyon (Williamson et al. 1977) seem clearly to have been designed for calendrical purposes.[13] We are not yet at the point where we can begin to differentiate between the astronomical practices of the Hovenweep Anasazi and the Chaco Canyon Anasazi, as we can differentiate their pottery types, for instance, or their building techniques. As we amass more data on the astronomical alignments created by the Anasazi and develop interpretations based on historical practices, however, we may eventually be able to discover the principles needed for such a differentiation.

Even after some 10 years of study, we are still in a nascent state with respect to Anasazi astronomy, but we can see considerable progress. For example, we can be much more confident now of Ellis's (1975) assertion that there is a historical continuum of Pueblo astronomical practices, and that early practices would be recognized today, and vice versa. We can now begin to hope that we can indeed use our interpretation of Anasazi astronomical practices to say more about other aspects of their ritual life. Clearly caution is needed, for many wild things have been suggested--that Casa Rinconada was used as an eclipse predictor (Morgan 1977), for example. Nonetheless, observations of the celestial sphere are an integral part of historical Pueblo life, and there is no reason to believe that these practices arose in recent times.

MORNING STAR, EVENING STAR

Four days they grew,
These two.
Then took the
Rainbow path
To claim
Their distant father.

Sun.

He of the above,
Their mother,
This world.
The two
Of Sun
And Earth.

Opposites reconciled.

Sun, for his part
Tested them fully,
Doubtful.
But
Seeing their mastery,
Pleased.

His sons.

Resilient lightning
He gave them.
Power,
To clear the monsters
From his world
And ours.

Two boys.

ACKNOWLEDGMENTS

I thank Jane Young and Michael Zeilik for their helpful comments on early drafts of this paper, and I appreciate the advice of Curt Schaafsma concerning the probable whereabouts of original excavation notes on Casa Rinconada. Part of this research was carried out with a National Geographic Foundation grant, and part was conducted through the research program of the Crow Canyon Campus of the Center for American Archaeology.

NOTES

1. The elements of this poem and its sequel at the end of this chapter are derived from several Pueblo origin myths, but primarily from Hopi and Zuni versions (Benedict 1935; Stephen 1929).

2. Historical analogy is a potentially powerful tool, especially in the Southwest, because many of the traditions manifest in the historical pueblos have clear antecedents in the prehistoric remains. It must be applied with considerable caution, however, because of the possibility that discontinuities in behavior have occurred, either as a result of independent invention or through outside influences.

3. As noted in Williamson (1982), comparisons between measurements made during excavation and recent transit measurements indicate that the differences in placement of the features of the reconstructed kiva and the original are insignificant.

4. In the present reconstruction of Casa Rinconada the distance between niches 27 and 28 (22 degrees 16 minutes) is nearly twice the average distance between niches. The field notes indicate evidence for a niche between 27 and 28 (Williamson 1982). It was not included in the reconstruction, probably because its precise location was uncertain.

5. It is impossible to determine from the archaeological remains whether the northeast opening is a window or a doorway. There is evidence that a room existed on the northeast side of the kiva at one time. If the room existed when the opening was built, it would probably have served as a doorway. Even then it would have been possible for the builders of the kiva to have incorporated the summer solstice alignment in the kiva by constructing an opening in the exterior wall of the room in line with the opening and niche E across the kiva.

6. By suitably narrowing the northeast port with plaster, or by using a narrow port in the exterior wall of the appended northeastern room, observers could have used the alignment for calendrical purposes. Given the obvious ritual use of the kiva and the other, noncalendrical orientations, however, I feel that the hypothesis that the summer solstice alignment was used for celebratory rather than calendrical purposes fits the available data better.

7. Ellis (1975) has argued persuasively that the Pueblo calendar is at least 1000 years old. Since it is virtually impossible in the Pueblo world to separate practices concerning the calendar from ritual practices, it seems reasonable that ceremonial procedures relating to the determination of calendrical events might be of similar antiquity.

8. Michael Zeilik has argued (in Chapter 3 of this volume) that it is unlikely that the northeast opening was used for this purpose or as a calendrical device because the northwest roof support would have blocked the line of sight to the lower niche opposite. If the post were large enough, this is possible. That argument only holds true, however, for the alignment between the opening and the niche. It may be that the intended alignment was actually between the northeast opening and the post. Such a hypothesis would fit nicely with the hypothesis that the posts were placed along cardinal directions for ritual reasons.

9. It is particularly interesting (and is a topic that perhaps deserves additional study) that the relationship between time

and space in a horizon calendar is nonlinear, i.e., the ratio between the passage of an interval of time and a unit of space is not constant and does not change at a constant rate.

10. Such a use of shadow casting is not unknown among historical groups. Ellis (1975:82) notes that near the summer solstice in one (unnamed) Tanoan pueblo "the head War Captain and the War Priest go to a solar monolith at the western edge of the pueblo. This is a white outcrop some two feet high and six inches thick, with an equal-armed cross representing a star cut into its east face. The shadows of the mountain range to the west are watched as they fall across this stone. If the shadow completely covers the stone, the omen is interpreted as foretelling a long summer; if the shadow is short, the summer will be correspondingly short and early frosts will ruin some late crops. The watchers at once hurry to announce the result of their observation to the religious society which is in retreat on this day and its members rejoice or lament according to the shadow's length."

11. Ellis (1975), citing others, has suggested that the Hopi watched sunrise to determine summer solstice and watched sunset to determine winter solstice. Zeilik (1983) has shown in his survey of the literature, however, that there seems to be no such preference at Hopi. Ellis's inference appears to have been an artifact of selective scanning of the literature.

12. For the summer solstice alignment of shadows to fall along the line between the northwest corner of the tower and the northeast corner of the roomblock, the north wall of the tower would have had to be located about 2.5 m beyond our best estimate of its placement. The shadow cast by the corner as we have estimated its location, however, could also have been used to mark the end point of the sun's yearly sweep along the horizon.

13. I suspect that this may be an accident of the structures we have studied and not a reflection of the overall astronomical practices in Chaco Canyon. It should be noted, however, that the structures in Hovenweep are between one and two centuries more recent than those in Chaco Canyon.

REFERENCES CITED

Benedict, Ruth F.
1935 Zuni Mythology. Columbia University Contributions to Anthropology 21. New York.

Chamberlain, Von Del
1982 When Stars Came Down to Earth. Ballena Press, Los Altos, California.

Curtis, Edward S.
1922 The North American Indian. Norwood, New York.

Cushing, Frank G.
1967 My Adventures at Zuni (reprint). Filter Press, Palmer Lake, Colorado.

Dorsey, George A.
n.d. Notes on file, Field Museum, Chicago.

Eliade, Mircea
1959 The Sacred and the Profane. Harcourt Brace, New York.

Ellis, Florence Hawley
1967 Where Did the Pueblo People Come from? El Palacio 73:35-43.
1975 A Thousand Years of the Sun-Moon-Star Calendar. In Archaeoastronomy in Pre-Columbian America, edited by Anthony F. Aveni, pp. 59-88. University of Texas Press, Austin.

Ellis, Florence Hawley, and Laurens Hammack
1968 The Inner Sanctum of Feather Cave, a Mogollon Sun and Earth Shrine Linking Mexico and the Southwest. American Antiquity 33:25-44.

Fewkes, J. Walter
1893 The Pa-lu-lu-kon-ti. Journal of American Folklore 6:269-282.

Frazier, Kendrick
1978 Solstice Watchers of Chaco. Science News 114:148-151.

Haile, Bernard F.
1947 Starlore among the Navaho. Museum of Navajo Ceremonial Art, Santa Fe.

Morgan, James
1977 Were Chaco's Great Kivas Ancient Computers of Astronomy? El Palacio 83:28-41.

O'Bryan, Aileen
1956 The Dine: Origin Myths of the Navaho Indians. Bureau of American Ethnology Bulletin 163:1-194. Smithsonian Institution, Washington, D. C.

Ortiz, Alfonso
1969 The Tewa World. University of Chicago Press, Chicago.

Ortiz, Alfonso
1972 Ritual Drama and the Pueblo World View. In New Perspectives on the Pueblos, edited by Alfonso Ortiz, pp. 135-157. University of New Mexico Press, Albuquerque.

Parsons, Elsie Clews
1925 A Pueblo Indian Journal. American Anthropological Association Memoirs 32:1-132.
1930 Isleta. Bureau of American Ethnology Annual Report 47:193-466. Smithsonian Institution, Washington, D. C.
1939 Pueblo Indian Religion. University of Chicago Press, Chicago.

Sofaer, Anna, Volker Zinser, and Rolf Sinclair
1980 A Unique Solar Marking Construct. Science 206:283-291.

Stephen, Alexander M.
1929 Hopi Tales. Journal of American Folklore 42:3-72.

Stirling, Matthew W.
1942 Origin Myth of Acoma. Bureau of American Ethnology Bulletin 135:1-123. Smithsonian Institution, Washington, D. C.

Vivian, Gordon, and Paul Reiter
1960 The Great Kivas of Chaco Canyon and Their Relationships. School of American Research and Museum of New Mexico Monographs No. 22. Santa Fe.

Williamson, Ray A.
1978 Pueblo Bonito and the Sun. Archaeoastronomy 1:5-6.
1982 Casa Rinconada, a Twelfth Century Anasazi Kiva. In Archaeoastronomy in the New World, edited by Anthony F. Aveni, pp. 205-219. Cambridge University Press, Cambridge, England.
1984 Living the Sky: The Native American Cosmos. Houghton Mifflin, Boston.

Williamson, Ray A., and M. Jane Young
1978 An Equinox Sun Petroglyph at Hovenweep National Monument. American Indian Rock Art 5:70-80.
1983 Cajon Field Report. Crow Canyon Campus, Center for American Archaeology. Cortez, Colorado.

Williamson, Ray A., Howard J. Fisher, and Donnel O'Flynn
1977 Anasazi Solar Observatories. In Native American Astronomy, edited by Anthony F. Aveni, pp. 203-218. University of Texas Press, Austin.

Winter, Joseph C., editor
1976 Hovenweep 1975. Anthropology Department, San Jose State University, San Jose, California.

Zeilik, Michael
1983 Historic Puebloan Sun Watching. Paper presented at the First International Ethnoastronomy Conference, Smithsonian Institution, Washington, D. C.

Chapter 8

Priests, Power, and Politics:
Some Implications of Socioceremonial Control

Jonathan E. Reyman

Anthropologists have studied the Pueblos intensively for more than a century beginning with the research of Bandelier, Cushing, Fewkes, Stephen, Stevenson, and others. With the use of Spanish and other records, our knowledge of the historical postcontact Pueblos can be extended back to the arrival of the first Europeans in the Southwest--the Spanish friar Marcos di Niza's entrada of 1539. Although the records are far from complete because much material was either accidentally or intentionally destroyed or lost (Kessell 1979; Lange 1982: 10), the extant published documents are rich in detailed ethnographic, ethnohistoric, and archaeological data.

It is likely that even more material remains unpublished, much of it in museums and archives. The National Anthropological Archives (an affiliate of the Smithsonian Institution, Washington, D. C.), for example, contain a great number of important unpublished notes, records, and manuscripts by Matilda Coxe Stevenson, John P. Harrington, and other early researchers; the papers of Neil M. Judd and Frank H. H. Roberts, Jr., have only recently been catalogued (Glenn 1982; Saquet 1983). These materials have not been studied adequately. I know of other early materials, including at least 10,000 pages of documents pertaining mostly to the Southwest, that are similarly unstudied and unpublished.

There are also materials that remain to be discovered. In February 1983, for example, I learned of and had the opportunity to examine a collection of archaeological materials from Hodge's 1917-1923 excavations at Hawikuh and accompanying ethnographic collections from Zuni. The owner told me that neither collection had been studied by an anthropologist, and I am now trying to obtain funds to study and prepare a catalog of these materials.

Anthropologists have been most interested in two major areas of Pueblo culture, the first being information on kinship, social structure, and social organization and the second being religion and ceremonial behavior. Many studies deal almost exclusively with these topics (e.g., Eggan 1950; Kroeber 1917; Parsons 1929, 1939), and even the more general ethnographies usually have considerable space devoted to these areas (e.g., Lange 1959; Parsons 1925b, 1936; Stevenson 1904; White 1932a, 1932b, 1942, 1962). It should be noted, however, that because social organization and religion pervade virtually all aspects of Pueblo life, even studies focused entirely on these topics necessarily deal, at least briefly, with matters of Pueblo ecology, politics, etc.

Despite this interest in Pueblo socioceremonial structure and behavior, comparatively little attention has been paid to the impact of the socioceremonial system on Pueblo life. Why this should be is unclear, but it may be a result of the assumption, both explicit (Benedict 1934; Colton 1934; Dozier 1954; Lange 1982; Titiev 1944) and implicit (Dozier 1970; Lange 1959), that the Pueblos are egalitarian societies. The following are typical statements about what has been called "the amorphous Hopi state" (Titiev 1944:59-68; see also Colton 1934):

> Functionally, the Hopi represent an egalitarian society with only minor authority vested in the village chief. By presenting problematic situations in a ritual context, they are confronted with the authority of convention and the judgment of laughter [Hieb 1979: 181].

> The practice of sharing food and work has also prevented social stratification; that is, there are no pronounced differences between poor and wealthy families [Dozier 1954:353].

If one assumes, a priori, that the Pueblos are egalitarian, then there is no reason to suppose that the behavior of a particular group within the pueblos, in this case the priests, either has implications for or could be the basis of ranking and social stratification. Furthermore, if one assumes the Pueblos to be egalitarian, then one probably would not look for evidence to the contrary and would not consider such evidence, if found, to be very significant. The above quotation from Dozier illustrates this last point. If, indeed, there are no pronounced differences, then on what basis did he distinguish rich from poor families? Ranking is implicit in the comparison. Of course, there could be differences between rich and poor without these differences being pronounced; however, the data, especially for the Hopi, indicate that the differences are significant, as documented later in this chapter.

I have previously dealt briefly with some nonegalitarian features of Pueblo societies (Reyman 1971, 1978, 1980). I will discuss these in greater depth here, although the discussion is still limited by considerations of length. The discussion is confined to published literature (and a few of my field notes) to emphasize my point in terms of the generally available data. As is shown, however, much of the same literature that describes the Pueblos as egalitarian also contains substantial data to support the hypothesis that they are nonegalitarian. These data suggest that all of the pueblos are rank societies and at least several are stratified. Upham (1982) makes a similar argument for the Western Pueblos, although McGuire (1983) criticizes Upham for confusing authority and power and for discussing Pueblo power in terms of control of ceremonial knowledge. Here I correct and add to Upham's data for the Western Pueblos, expand the discussion to include the Eastern Pueblos, consider some material correlates of ceremonial knowledge to meet McGuire's objections, and present some data to support further Upham's argument regarding the post-Pueblo IV attenuation of Pueblo culture (cf. Reyman 1971, 1976b; McCluskey 1977, 1981, 1982). Finally, I consider some implications of this discussion for the Chacoan Anasazi.

SOME CHARACTERISTICS OF RANK SOCIETY

"An egalitarian society is one in which there are as many positions of prestige in any given sex-age grade as there are persons capable of filling them" (Fried 1967:33). Economically, an egalitarian society is characterized by a reciprocal economy in which the members have similar access to valued resources (Fried 1967:35).

By contrast, a nonegalitarian or rank society "is one in which positions of valued status are somehow limited so that not all those of sufficient talent to occupy such statuses actually achieve them. Such a society may or may not be stratified" (Fried 1967:109). Rank societies have been characterized as having a redistributive economy, a village-based settlement pattern, and political and ritual leadership without effective sanctions to compel obedience. Ritual leadership is often vested in the same individuals as political leadership (Fried 1967:116-141). In rank society, leaders have authority--"the ability to channel the behavior of others by the threat or use of sanctions" (Fried 1967:13). [Note: Fried and others have suggested a number of other characteristics of rank society that cannot be considered here for lack of space.]

A stratified society is one in which the characteristics of a rank society have been developed even further. "A stratified society is one in which members of the same sex and equivalent age status do not have equal access to the basic resources that sustain life" (Fried 1967:186). Furthermore, leaders assign "all available usufructs to specific individuals or groups, the latter being composed of members fewer than the total population of the society" (Fried 1967:188). Finally (for my purpose, at least, since again there is not enough space to discuss all the relevant characteristics), leadership is hereditary and is maintained through a system of conventional behavior (e.g., endogamy and inheritance rules) and sanctions. The latter range from relatively mild to severe, the degree of severity being dependent upon the specific behavior of the transgressor and its consequences for the group. Leaders in stratified societies generally have both authority and power, as defined above.

PUEBLOS AS NONEGALITARIAN SOCIETIES

A confirmation of the hypothesis that the Pueblos are nonegalitarian (rank) societies requires that evidence be provided for the various characteristics noted in the preceding section. At least two of the characteristics are generally agreed upon: the Pueblos have (a) a village settlement pattern and (b) a redistributive economy. Some of the best evidence for the latter comes from the Western Pueblo kachina ceremonies, especially the public rites, where large quantities of food and other items (traditionally including baskets, woven plaques, and kachina dolls) are collected at some location and then distributed publicly during breaks in the dancing. Public redistribution also occurs at other ceremonial occasions among both Eastern and Western Pueblos, including feast days, nonmasked dances (such as animal dances), and marriages (Beaglehole 1937; Eggan 1950; Lange 1959, 1982; Ortiz 1969; Parsons 1929, 1932, 1936, 1939; Stevenson 1904; Titiev 1944, 1972; White 1932a, 1932b, 1942, 1962).

The rest of this section will focus on demonstrating that the Pueblos fulfill the characteristics of rank and stratified societies, as defined by Fried. The Pueblos will be shown to have hereditary leadership, maintained through a system of conventional behavior, with varying degrees of sanction. Pueblo leaders, in general, will be shown as possessing both authority and power.

Rank and Social Stratification Among the Pueblos

Upham (1982:14-16) provides a good summary of the Hopi data from Eggan (1950), Parsons (1939), Stephen (Parsons 1936), and Titiev (1944) for the existence of three social classes or strata within the villages (cf. Dozier 1954), so I need only reiterate two major points. First, the basis for the three classes is found in the origin myths:

> The people were governed by the chiefs (Mongwi), village criers (Chakmongwi), priests (Momwit), and high priests and all their religious rites were ruled by the high priests. The people were classed as common, middle, and first class [Colton 1936: 7].

A note referring to the passage just quoted identifies the members of these three classes as follows:

> First Class, Mong-cinum--Leaders of the kivas, priests, and High Priests.
>
> Middle Class, Pavun-cinum--Hold no office but belong to societies and take part in ceremonies.
>
> Low Class, Sukavung-cinum--Do not belong to any societies nor take part in ceremonies.
>
> These classes refer only to men. When a woman marries, she is classed with her husband [Colton 1936:125].

The second point to be emphasized about Hopi class structure is that the origin myths also are the basis for the predominance of the Bear clan among the Hopi. The Bear clan was the first to arrive at the mesas and thereby gained control of all village lands and all major ceremonial and political offices (Eggan 1950:64; Parsons 1936:744, 1088; Titiev 1944, especially pp. 61 and 73-74; cf. Mindeleff [1891:17-18], who identifies the Snake clan as the first to arrive, and Titiev [1944: 79] regarding claims of the Spider clan). Thus clans are ranked at Hopi, forming a hierarchy that is reinforced by the grouping of high-ranked clans into phratries (Eggan 1950; Titiev 1944).

Although clans are exogamous, an analysis of marriage patterns using data from Titiev (1944) and Eggan (1950) indicates a high degree of phratry endogamy. This marriage pattern perpetuates the three-level system of stratification and helps to consolidate sociopolitical and ceremonial control within the kin groups that hold offices. Contrary to Upham (1982:14-16), it is not the clan that is the basic unit for hereditary transfer of the important priestly offices; rather, specific lineages within the particular clans control these offices.

> The closest and most important set of relatives make up the lineage group, and within this group occur both the most varied obligations and ties and the greatest differentiations in terminology [Eggan 1950:60].
>
> The most important structural principal among the Hopi, however, is . . . the solidarity and unity of the lineage group [Eggan 1950:112].
>
> A significant aspect of the lineage . . . is its function of conserving and transmitting

> not only the social and ritual heritage, but property rights, houses, and ceremonies, as well, from one generation to the next [Eggan 1950:114; emphasis added].

A similar class structure prevails at other pueblos. Again, Upham (1982:16-19) summarizes the data for Zuni and Acoma, so there is no need to repeat them here except to reaffirm and amplify two points of similarity with the Hopi. First, at Acoma, as at Hopi, it is a specific lineage within the Antelope clan that controls the office of *ha'actitcani*, "a person who symbolizes or represents the whole pueblo" (White 1932a:41), or town chief, the priest we usually call the cacique. "Acoma's Town Chief is chosen within the Antelope clan and, as far as information goes, within one lineage, by Antelope clansmen" (Parsons 1939:882). At Zuni, the Dogwood clan traditionally controlled the two major offices, the Chief Priest (*shi′wanni*) of the North (the *kia′kwemosi*) and the *pekwin* (spokesman priest or priest of the zenith, usually called sun priest in English), as well as providing the heads and members of other major priestly orders (Eggan 1950:220-221; Ladd 1979; Stevenson 1904). Furthermore, as Stevenson states,

> The priesthood is confined to families, the rule being that each member of a division of a priesthood must be of the clan or a child of the clan of the *shi′wanni* of the division. The son or brother of the *shi′wanni* fills a vacancy, *preferences being given to the eldest son* [Stevenson 1904:165; emphasis added].

Thus, control is held by specific lineages within particular clans where possible.

A second similarity to the Hopi case is that

> In Zuni definition, poor people are people without ceremonial property or connection, belonging to no rain or curing society--people who are not "valuable." In this sense, the poor of Zuni are about half the population. The proportion of poor would be very much larger in the East, but I doubt if the distinction between ceremonially poor and rich would be made at all there;* the membership lists are too small. . . .
>
> *Among the Keres there is a term, *sishti*, denoting any person not actually engaged in ceremonial work (Santo Domingo) or any common persons without ceremonial (or governmental) affiliation [Parsons 1939:112; for an alternate view of the Eastern Pueblos, see below].

A review of the data for Acoma (White 1932a), my own research at that pueblo, and a comparison with other Keresan pueblos indicate that the situation at Acoma is similar to the Hopi and Zuni cases. There is a basic division between those with ceremonial offices or duties (high rank) and those without such responsibilities (low rank). For example, Lange (1959, especially pages 269-270 and 284-286) notes that the Cochiti refer to nonsociety members as *sir′shtī* (*crudos* or "raw") as opposed to the Ku-sha′lī and Kwe′-rana, who are "cooked." There is some inconsistency because the cacique is *not* supposed to be a society member but rather should be a *sir′-shtī*, "one of the general public" (Lange 1959:239). Nevertheless, as Lange (1959:237-252, 457) points out, caciques have usually been members of the major societies, especially Flint and Ku-sha′lī.

Among the eastern Keres, the rules vary as to which kin groups control the office of cacique. At Zia, the cacique should be chosen from one of five specific clans, and the office should rotate among them; a tiamunyi (cacique) from one clan should not be succeeded by another of the same clan (White 1962:127). At Santa Ana, there does not seem to be a specific rule, or at least none is reported (White 1942). At Cochiti, the cacique is not supposed to select his successor from his own family. In one case, a father and son both served as cacique, but not in succession (Lange 1959:241-242).

At Jemez, the cacique and his two assistants must come from either the Young Corn or the Sun clan but not from any society membership (Parsons 1925b:59). Among the Tewa, "inheritance of office [cacique or town chief, of which there are two, one for the summer moiety and one for the winter moiety] through kinship is entirely out of the question, theoretically; in practice there may be a tendency to choose kinsmen as assistants" (Parsons 1929:109). As one of Parsons's informants from San Ildefonso noted, "When there is a relative they put him in, so they won't have to come out from their house" (Parsons 1929:109).

The Tewa also distinguish the Dry Food people (common Tewa who serve no official capacity in the political or ritual system) from the towa′e (six native political officers chosen to serve one year from among the Dry Food people, the group to which they return after their year of service) and the Made people (society members). Dry Food people are also called the Whe Towa (weed or trash people) and Nayi wha Towa (dust-dragging people), the "latter term deriving from a prior practice whereby the common people swept the village before ritual and other festive occasions" (Ortiz 1969:17).

These three levels of existence correspond to three levels of being in the spirit world and are defined by Tewa myth and cosmology (Ortiz 1969). The eight groups of Made people constitute a hierarchy by which is meant that, "with the exception of the Hunt society, this is the order in which they were 'made' in the myth of origin" (Ortiz 1969:83). Once again we have a native system of stratification in which the two caciques alternate control, the changeover coming (in theory) at the equinoxes (Ortiz 1969:104-119; cf. Hill in Lange 1982).

Contrary to Parsons (1939:112), "the Made People and their assistants comprise a very sizable percentage of the population" (Ortiz 1969:82), although certainly not close to a majority. There is a general attitude that one ought not remain a Dry Food person all one's life (Ortiz 1969:48). To be a Made person is to occupy a higher status in terms of both human existence and the spiritual world.

It is clear, therefore, that the Pueblos recognize differences in social statuses and that these differences are hierarchically ranked. This ranking is somewhat better defined among the Western Pueblos than among the Eastern Pueblos, but it is present in both areas. High rank is everywhere dependent upon ceremonial knowledge or office, which in turn is usually acquired through specific lines of descent. This hereditary system for succession is also better demonstrated in the west, but at least vestiges of a similar system remain among some of the Eastern Pueblos (for a general summary, although one with some errors and omissions, see Parsons 1939:Table 1).

To a significant extent, the system is closed. That is, to become a high-ranked officer, especially in the Western Pueblos, one must have the requisite ceremonial knowledge, but unless one is a member of a particular lineage and clan, one will never be in a position to acquire this knowledge. Descent is the basic criterion, along with the added factors of ceremonial knowledge, age, personal qualities, etc., in determining who will succeed to a given office.

> Obviously the selection and education of those who may become chiefs is an important matter. The selection does not go strictly by birth or seniority or other form of automatic succession even when these are prevailing principles of choice. Character, interest, and mentality, particularly memory, are taken into account. To be a good chief, a man must be self-controlled, patient, enduring, and devoted; to have "power," he must know long and difficult prayers or chants. In council and sometimes before the people he must be skilled in the art of oratory. A Hopi chief has to choose his successor within his material _lineage_, theoretically his successor is his younger brother or the eldest son of his eldest sister, but practically a chief may choose a junior nephew if he think[s] the youth is better qualified [Parsons 1939:154; emphasis added].

Sometimes the office and its duties may be transferred to another priest if the incumbent is removed or if he dies with no legitimate successor. The assumption of the pekwin's duties by the _kia'kwemosi_ (Chief Priest of the North) at Zuni is an example of this. Alternately, an office and its responsibilities, even entire societies, sometimes become extinct as incumbents die without training successors, as happened with the Snake, Fire, and other societies at Cochiti (Lange 1959:265-266, Appendix 17). Thus it is clear that the Pueblos meet the criterion for nonegalitarian societies of restricted access to positions of valued status.

Leadership Positions

Two basic concepts regarding leadership and the authority and power associated with it run through Pueblo society. The first is that one should not deliberately seek a leadership position, or at least should not appear to seek it:

> The man, above all, who thirsts for power or knowledge, who wishes to be, as they scornfully phrase it, "a leader of his people," receives nothing but censure and will very likely be persecuted for sorcery [Bunzel 1932:480].

At Santa Ana, it was reported that reluctance to accept and confirm a particular man as cacique derived from the fact that the man "put himself up" for the position (White 1942:95-96). Exceptions to this norm of behavior do occur. At Taos, a man became a kiva leader through his aggressive pursuit of the position; although many in Taos resented his open drive for and assumption of authority and power, this resentment did not prevent him from seizing control (Smith 1969:26).

The other basic concept of leadership dictates that when one is chosen for office and is installed in it, one is to work for the good of the pueblo and not for personal gain. The higher the office in the priestly hierarchy, the more this pertains. The rain priests (_shi'wanni_) at Zuni and the cacique among the Keres theoretically are not supposed to have any involvement in mundane affairs or even to hear quarrels and other indicators of dissension, lest they be distracted in some way from their prayers and other ritual behaviors on behalf of the village (Eggan 1950:204; Lange 1959; White 1932a, 1932b, 1942, 1962).

The number of leadership positions is severely limited for all the Pueblos. Among the Hopi, there is only one town chief (kikmongwi), one sun priest (ta'wa mongwi), one chief of the Powamu society, and so forth. At Zuni, there was one pekwin, one kia'kwemosi, an elder and a younger bow priest (plus assistants), and a very limited number of other priests within the fourteen rain (shi'wanni) priesthoods (cf. Eggan 1950; Ladd 1979). At Acoma, Laguna, and the eastern Keresan pueblos, there is one cacique (Santa Ana, an exception, has two kinds of caciques [White 1942: 98]), one head of each medicine society, one war priest (nahī'ya, a position second in the hierarchy to the cacique and now extinct at some pueblos--Cochiti, for example [Lange 1959:276]), and a single head for each "managing society," e.g., Kusha'lī and Kwe'rana. Jemez has a cacique, two assistants, and a limited number of other priests (Parsons 1925b). The Tewa, as noted, have dual leadership vested in a pair of moiety chiefs and a hierarchy of eight groups of Made people. Although a significant number of people belong to these eight groups, the leadership positions within them are tightly restricted (cf. Lange 1982; Ortiz 1969; Parsons 1929). The Tiwa, as represented by Isleta, for example, have a single cacique (t'aikabede) and seventeen ceremonial groups, most of which have one head priest and one assistant plus a limited number of members (Parsons 1932:255-256).

This information and the observation that, until recently, trained successors were ready to assume each office, indicate that another criterion for a ranked society is met among the Pueblos: positions of valued status are limited in number so that there are more persons of sufficient talent than there are positions available, and not all who are qualified to fulfill an office achieve it.

The Basis for Priestly Authority and Power

Upham (1982:19-20) argues that priestly authority and power rest not "on personal material wealth but on access to, and possession of, ritual and ceremonial knowledge," which in turn are vested in the clan and protected by a system of "hereditary oligarchy." Following Brandt (1980), he argues that access to this knowledge is restricted and maintained through institutionalized secrecy. McGuire's (1983) objections to Upham's argument were cited earlier, and I have shown that, especially for the Western Pueblos, it is the lineage and not the clan that is fundamental to the transmission of high-ranking priestly offices.

Authority and power are not synonymous, as discussed above. I agree with McGuire that ritual knowledge alone is a weak basis for either, but I am not certain that, as he contends, Upham confuses the two. Rather, I would suggest that Upham has not carried his argument far enough and has not fully considered the implications of priestly control of ritual knowledge. He also has not examined closely enough the ethnographic data on material correlates to priestly power. Priestly authority and, to a greater extent, priestly power should have significant material correlates. If Pueblo societies are nonegalitarian (rank or stratified), then it is likely that hierarchical status differences based on differential access to ceremonial knowledge and positions result in differential access to and control over valued resources.

The ritual and ceremonial bases for priestly authority and power are discussed below. Consideration of the material correlates of priestly office follows this discussion of ritual and ceremonialism.

Generally, the priest who "keeps the calendar" controls pueblo activities in a very real sense and often can be regarded as the most important official in the village regardless of whether or not he is the town chief, per se; at the least, he is a coequal with the town chief where the positions are vested in different individuals, notably among the Hopi and Zuni. Among the Hopi, the *ta'wa mongwi* (sun chief or sunwatcher), not the *kikmongwi* (town chief), keeps the calendar (Parsons 1936; Simmons 1942; Titiev 1938, 1944). At Zuni, the pekwin kept the calendar until the 1940s (Ladd 1979), at which time the Chief Priest of the North (*kia'kwemosi*) took over the sun watching and other duties. At Acoma and among the Eastern Pueblos, the cacique usually keeps the calendar. Because I have discussed Pueblo astronomy at length elsewhere (Reyman 1971, 1975, 1976a, 1976b, 1979, 1980, 1982), only the most basic features are reviewed here.

The calendar is based on astronomical observations, most importantly solar observations. This is reflected in the Hopi term for their astronomer-priest (*ta'wa mongwi*) and in the common translation of "sun priest" for pekwin. Lunar and stellar observations are more important among the Western than among the Eastern Pueblos (cf. Bunzel 1932; Parsons 1936; Reyman 1980), which probably reflects the more precarious environmental conditions found in the west. Even in the west, lunar observations are decidedly secondary to solar observations, and stellar observations, although useful, are still less important.

Because the sun priests make the observations and keep the calendars by which Pueblo socioeconomic and ceremonial behaviors are planned and implemented, their ability to perform their duties is critical to the survival of the village (although more so from a historical perspective than today). Sun priests have to be reasonably accurate in their observations, record keeping, and predictions of such astronomical events as solstices and equinoxes.

> The observation of these sunrise positions is . . . skilled work, and it is the duty of the religious official, the Sunwatcher, to observe them, to forewarn the people of important dates, and to announce them in due course [Forde 1931:384].

> . . . it may be that the time set for "calling the sun back" intentionally preceded the true solstice in order that it could be claimed that the recall was actually achieved. Similarly, late in the fall, after the main harvests, the reversal of the sun's course along the eastern horizon might have been intentionally anticipated in an effort to impress the laymen with the powers of the medicine societies [Lange 1959:321-322; see Reyman 1980 and Zeilik (Chapter 3, this volume) for further discussion of the "anticipation" issue in astronomical observations and calculations].

The importance of observational skill is also noted in the following statement from *Zuni Mythology*:

> The man who went to the Sun was made Pekwin. The Sun told him, "When you get home you will be Pekwin and I will be your father. Make meal offerings to me. Come to the edge

of town every morning and pray to me. Every evening go to the shrine at Matsaka and pray. At the end of the year when I come to the south, watch me closely; and in the middle of the year in the same month, when I reach the farthest point on the right hand, watch me closely." "All right." He came home and learned for three years, and he was made Pekwin. The first year at the last month of the year he watched the Sun closely, but his calculations [for the winter solstice] were early by thirteen days. Next year he was early by twenty days. He studied again. The next year his calculations were two days late. In eight years he was able to time the turning of the Sun exactly [Benedict 1935(II): 66-67].

In these terms, priestly power rests, to a significant extent, on what I have called the "predictive dimension" (Reyman 1980; see also Parsons 1939:84-89), the ability to predict seasonal changes in order to plan and execute the proper sequence of subsistence activities and the accompanying requisite ceremonies. Knowledge of the major solar events; calculations of some intermediate dates associated with seed preparation, irrigation ditch maintenance, planting times, etc. (Forde 1963: 227; Ortiz 1969:Figs. 9-11; Parsons 1936:Map 4; Titiev 1938); and observation of changes in flora (Ellis and Hammack 1968:30-31) and fauna probably provided as much predictive accuracy in terms of climatic conditions as is possible in the Southwest in the absence of long-term written records (McCluskey 1982; Reyman 1976a, 1980, 1982; Zeilik, Chapter 3, this volume).

Lunar observations and record keeping add relatively little to predictive accuracy, although where made (notably at Hopi and Zuni), they may be useful for scheduling some planting activities (Parsons 1936:1037-1041) and for predicting animal migrations in conjunction with hunting activities. Although the moon is everywhere considered a deity, if only a minor one, some pueblos, such as Cochiti, neither observe it regularly nor make lunar counts for reckoning months (Lange 1959:322). The moon is similarly unimportant for astronomical record keeping among the Tewa (Ortiz 1969) and the Keresan pueblos (Goldfrank 1927; White 1932a, 1932b, 1942, 1962). There is no evidence that the Pueblos have any interest in observing or recording the major and minor lunar standstills, nor are they interested in eclipse prediction (for which standstill records might be useful). There is some fear of eclipses (Parsons 1932:342, 1936:984, 1939:86), but as two Hopi told me, although eclipses are dangerous times, they are not very important because they pass quickly. The lack of ethnographic data regarding lunar observations, especially of standstills, casts doubt on many of the conclusions of Sofaer and her colleagues (Sofaer et al. 1982; Sofaer and Sinclair, Chapter 4, this volume).

Stellar observations usually serve as nighttime checks on the sun's position, especially when observations of sunrise and sunset are obscured by clouds or other adverse weather conditions. These observations are also used for timing the sequence of events during particular ceremonies. Stellar observations are more commonly used among the Western Pueblos, most notably the Hopi where the stars of Orion's belt and the Pleiades are the most frequently used timing stars (Parsons 1936). The heliacal rising of Venus is also used to mark the beginning or end of specific events during certain ceremonies (Cushing 1882b: 508; Parsons 1936).

We must be careful not to impute too great a degree of accuracy to Pueblo astronomy, a caution supported by the ethnographic records. Pueblo astronomers do make mistakes, as the above quotation from Zuni Mythology indicates. Titiev (1938) cites a case at Hotevilla where the priest miscounted an entire moon, and Parsons (1925a:75) records several errors by the ta'wa mongwi at one pueblo. Although Pueblo astronomy was probably derived, to a significant extent, from various Mesoamerican systems (Reyman 1971, 1976b), we are not dealing with Maya priests at a more northerly latitude; Pueblo astronomy did not reach the level of refinement and accuracy found in Mesoamerica.

Some of the priests' ritual knowledge is esoteric and secret (Brandt 1980; Lange 1959:249, 321-322; Ortiz 1969; White 1962:228), and priests who talk about their ritual knowledge lose their power: "power talked about is power lost" (Parsons 1939:83). Few people, if any, know exactly what priests do in carrying out the functions of their offices. Successors must be taught the proper ritual behavior, including performance sequences, and this training takes years to complete (if, in fact, it is ever finished). The secrecy surrounding ritual knowledge explains why, in some cases, offices or societies become extinct; there would be no extinction if ritual knowledge were common property. In cases where the priest's duties are essential to the pueblo's survival, however, there are apparently provisions for others to assume the responsibilities, as at Zuni where the office of pekwin was vacated in the 1940s (Ladd 1979). Informants from an Eastern Pueblo have told me of a case where an entire society was reinstituted several years after its extinction with the death of the last member. There is obviously less secrecy among the priests than between the priests and the general populace.

Much of what the priests do, and perhaps how they do it, is public knowledge. In the two Hopi examples noted above (Parsons 1925a:75; Titiev 1938), the people were well aware that the priests had made mistakes in their astronomical observations and calculations. "All of these points [horizon sunrise points marking planting dates] are announced by the Sunwatcher and many of these are generally known" (Forde 1931:385; see also Cushing 1883:39 for additional observations concerning Zuni).

On a more general level, the identities of the kachinas are secrets only in terms of uninitiated children, even though no one publicly professes knowledge of this "secret." "When the Katcinas entered the kivas without masks, I had a great surprise. They were not spirits but human beings . . . the head man told us . . . that if we ever talked about this to uninitiated children we would get a thrashing" (Simmons 1942:84). Nevertheless, secrecy is an additional factor, and an important one, by which restricted access to high-rank positions is maintained.

The institutionalization of secrecy also helps us to explain why, when a village divides as a result of a factional dispute (e.g., as at Oraibi), it is usually the conservative or traditionalist faction that leaves; it alone has the basis (secret knowledge) for reconstituting the community and continuing the customary lifeway (cf. Brandt 1980; Simmons 1942; Titiev 1944, 1972). Those who remain in the village often cannot maintain it ceremonially or otherwise, as the decline of Oraibi after 1906 indicates.

Access to and possession of ceremonial knowledge is restricted by descent, institutionalized secrecy, and the other factors noted earlier, all of which contribute to restricting the number of persons who can occupy the limited number of high-status offices. This further supports the hypothesis that the Pueblos are nonegalitarian societies.

The Material Correlates of Priestly Authority and Power

During the discussion that followed the original symposium presentation on which this paper is based, most comments and criticism were directed toward the portion of the argument that follows. The Zuni and Hopi attending the symposium raised two main objections. The first was that priests are responsible to the pueblo as a whole, and their behavior must be for the good of the village, not for personal gain. Above all, priests must be "good men" with "good hearts." The second objection was that priests do not have the power to command; "We are not cogs in a machine controlled by the sun priest," as one Zuni participant stated.

I tried to make it clear at the time (perhaps without much success) that I did not disagree with them, at least in part. Priests are supposed to have good hearts and to work for the good of the entire village. This is well documented in the literature: "the character of the chiefs should be above suspicion and a man's behavior affects the validity of his ceremony" (Parsons 1939:154). "Usually, distinction or power is achieved through social solidarity rather than through personal experience" (Parsons 1939:108). "Snake chief says that he wants to call out as soon as the sun gets to the place. So he has to watch very closely. He wants to do what is right. He wants to do his best" (Parsons 1925a:101; see also Parsons 1925a:75, Titiev 1938:39).

The ethnographic literature also provides ample evidence that priests do receive significant material benefits for their work, as well as nonmaterial rewards. Furthermore, as a consequence of their responsibilities and behavior, priests control a major portion of the energy flow within the villages. In reviewing the data discussed below and the comments made at the symposium, I have concluded that the objections to the materialist aspect of this argument probably reflect differences between an emic (Zuni and Hopi) and an etic view of Pueblo priesthood. These two views are more complementary than contradictory, however, as the following should help to make clear.

Although Upham (1982:16-18) notes that Hopi and Zuni priests have some access to and control over communal labor, as well as other material benefits, he argues that Pueblo social stratification is better understood in terms of control over ritual knowledge (Upham 1982:20). A review of the ethnographic data suggests that Upham underestimates the material benefits of holding priestly office.

The first and most obvious benefit is that the town chief and sometimes other priests, such as the pekwin, are relieved of all subsistence activities; their fields are planted for them, and other food is provided by communal labor (Beaglehole 1937:38; Hoebel 1969:99, 110; Lange 1959:83-85; Parsons 1925b:61, 1932:257, 1939:157-158; Titiev 1944:65; White 1942:97, 1962:125). There is some disagreement over the situation at Acoma and among the Tewa. Parsons (1929:110, 1939:157, 882) states that the caciques in these pueblos do their own farming. White (1932a:42) states that, at Acoma, the cacique's fields are planted for him under the direction of the war chief (which is consistent with practices reported for other Keresan pueblos), and Hill (in Lange 1982:25) indicates that the same is true for the Tewa village of Santa Clara.

Other basic services to the chief mentioned in these same accounts include cooking his meals; plastering the walls of his house; providing firewood, water, and other essential materials; and holding communal hunts, especially rabbit hunts, to provide him and his family with meat. As Parsons (1939:158) notes, "the chiefs work for the people at large, and the people work gratefully to pay them." Thus, the chiefs are the recipients of a great deal of energy expended on their behalf, regardless of whether or not they considered this upon assuming office and whether or not the people recognize it as payment, per se. Priests are compensated for their work.

The town chief often keeps and has access to the pueblo's surplus production and seed reserves. Most pueblos keep one or two years' supply in reserve. In times of shortage, the chief in charge of the communal stores has the greatest and most immediate access to this food (e.g., Parsons 1939:157-158, 882). He is able to allocate more seed or the best quality seed to his kin, starting with his own lineage. Informants of mine at two Eastern Pueblos complained that the caciques kept the village stores of seed in their "offices" and did not always give seed equally to whoever asked for it, as they are supposed to do (cf. Hoebel 1969:99). Recalling a period of scarcity, one Hopi noted, "Even if a Chief had a lot of corn he would not share it" (Udall 1969:44).

The town chief also has control over the assignment of land and water rights for the pueblo, although the assignments may be issued through lower-ranked officials. This means that he can allocate the best farmland to his kin, first to members of his own lineage and then to other clan members. At Hopi, the kikmongwi has a large plot of land set aside for his use, and he can use this land to reward those who support him in his ceremonial duties (Titiev 1944:62; Udall 1969:38-39). Coupled with his control of the surplus seed, this privileged access to agricultural resources increases the likelihood that his kin group will achieve farming success and high productivity more often than other groups within the pueblo. This, in turn, suggests that when surplus wealth is produced, these chiefs and their lineages are likely to control a good portion of it.

In time, if such conditions persist, one or more lineages are able to gain economic dominance over the others. When these lineages are further united through clanship, the result is a villagewide system of stratified lineages and clans. Linkage of clans through phratries can further this stratification. I suggest that this is what happened at Hopi and, perhaps to a lesser extent, at Zuni and elsewhere.

Theoretically, the town chief is supposed to act in the best interests of the pueblo as a whole. That he generally does is not in question, but the ethnographic data indicate that this is not always the case. Indeed, a dispute over the reassignment of lands by the kikmongwi to members of his clan (Bear) at a time of general economic hardship caused by long-term drought and arroyo cutting when Bear clan lands were being severely damaged by erosion was the underlying cause for the split of Oraibi in 1906 (Titiev 1944:60-64, 69-95; cf. Bradfield 1971:45).

Although it is contrary to Pueblo ideology, self-interest is clearly the motive for actions at times, and this results in the differential accumulation of wealth and access to resources, a nonegalitarian situation. In some cases, this may be more an accidental consequence of holding office than a deliberate attempt to become wealthy, but the situation occurs nonetheless. Furthermore, in the assignment of land and water usufructs by the town chiefs to specific individuals and groups, we find that the Pueblos satisfy another major criterion for stratified societies.

Whoever keeps the calendar, be it the town chief, ta'wa mongwi, or pekwin, is responsible for structuring pueblo life on both a day-to-day and a yearly basis. Almost all pueblo activities, and certainly all those that are critical to village survival, both mundane and sacred, are predicted, scheduled, and executed according to the calendar, e.g., seed preparation, ditch maintenance, planting, harvesting, hunting, etc., and the accompanying ceremonies. The cacique's approval must often be obtained in order to hold ceremonies controlled by the medicine societies (White 1932a:41). At Acoma and elsewhere, the town chief or other priest also decides which kiva groups or societies will participate in various ceremonies and when public curing ceremonies will be held (Lange 1959; Ortiz 1969; Titiev 1944, 1972; White 1932a, 1932b, 1942, 1962).

These announcements are not taken lightly. Pueblo people may not be cogs in a machine, but as Forde (1931:388) notes, some of the entries in A Pueblo Indian Journal (Parsons 1925a) suggest that when the ta'wa mongwi announces the dates for planting, these are orders to certain individuals or groups to plant.

Through control of the calendar, and especially the concomitant ceremonies, the sunwatcher and other priests stimulate economic production. The following account describes the situation among the Hopi, but similar data are available for the other pueblos (Lange 1959, 1982; Ortiz 1969; Parsons 1925b, 1932, 1939; Stevenson 1904; White 1932a, 1932b, 1942, 1962).

> All types of Hopi religious ceremonial react strongly on the functioning of economic patterns in that they provide a potent incentive to the increased production and distribution of native wealth. . . . For society ceremonials and for calendrical katsina dances, food and other provisions for the performers and their associates are provided by the cooperation of the household groups to which participants belong. . . .
>
> Preparations for this commence early in spring. . . . Thereafter much corn and other food must be accumulated and held in reserve for festival use. As soon as the days on which the ceremony is to be held are announced by the crier chief, the working rhythm of the village begins to move at a faster tempo. The men start work on their costumes . . . some set to work also to make new clothes for members of their household. . . . The men collect and carry much wood and water to the village. . . . The women grind corn and prepare large quantities of . . . food. Men and women combine to renovate houses thoroughly. . .
>
> The day of the dance marks the climax of this effort through the distribution and consumption of the goods which it has required so much labor

to prepare. Each household contributes its quota of food for the dancers. Large quantities are given to this end, the unconsumed food being divided equally among the dancers after the day is done, and much more is used to feast the household. . . . In one very definite sense, the performance of the ritual is secondary to the associated economic activity for which it is the occasion...

The ceremony over, and the prayers of the people sent forth, the Hopi can turn again to his work with a new meaning and interest strengthened by a feeling and vision . . . of the total social and religious content of which the work is an integral part. . . .

Some rituals leave behind them a species of economic debt, payment of which further increases the distribution of wealth. . . . In the case of all these ceremonial debts, return payment is made as soon as possible after the ceremony. This means that after a short rest period the recipients or members of their households begin to make baskets, grind corn, prepare foods or go hunting. . . .

Gift exchange and the payment of forfeits in winter game activities both help to further the circulation of goods in the community. Examples of the exchange of goods marking kinship obligations, of services performed in religious or personal ceremonial, and in the give and take of cooperative social life have already been mentioned in this and preceding sections [Beaglehole 1937, especially pp. 72-77]. . . . The concept of the free gift is one that is alien to the dominant economic patterns of Hopi society. No gift is ever made nor services rendered but that the culture provides definite customary roles for fixing the value of the return gift or payment and the occasion when it should be made [Beaglehole 1937:78-80].

Beaglehole (1937:72-77) also reports that increased economic production accompanies other ceremonies, especially those connected with the rites of passage--birth and naming, initiation, and marriage (death is not an occasion for a significant increase in production because the rites are brief and simple). Priests often have major roles in the rites, especially those of initiation into ceremonial societies.

The town chief or other head priest often appoints other pueblo officials, or at least approves their appointments. This authority extends to both sacred and secular offices, the latter group a relatively recent development in response to European intrusion (cf. Hoebel 1969:99). Among the Hopi, for example, the kikmongwi is able to appoint to positions of authority and power those persons who respond favorably to him and to his conduct of his office, and he is able to reward them with land and other benefits in exchange for their support (Titiev 1944:62). Even though the Hopi now have an elected tribal council, each kikmongwi must approve his village's representatives to the council (several 1984 issues of Qua' Toqti, the Hopi newspaper, have discussions of this situation--e.g., January 26, February 2, and February 16, 1984).

The town chief conducts the initiations of other priests and secular officials. Depending upon the pueblo, he or another priest presides over the initiation of the children into the kachina and other societies (Lange 1959, 1982; Ortiz 1969; Parsons 1925b, 1932, 1936, 1939; Simmons 1942; Stevenson 1904; Titiev 1944; White 1932a, 1932b, 1942, 1962). Thus, priests directly control admission to their ranks and dispersal of ceremonial knowledge, which further restricts access to high-ranked positions. Their authority to appoint others to positions of authority and power also means that both ritual and political leadership are vested in the same individuals to a significant extent, an additional argument for classifying the Pueblos among rank and stratified societies.

In summary, the town chief and other priests are the direct recipients of substantial material (economic) benefits, and significant energy is expended on their behalf by the rest of the villagers. Through control of the calendar they schedule, organize, and direct much of the communal behavior, thereby controlling a major portion of the energy flow within the pueblo. Because priests directly or indirectly control this wealth and labor (energy), they are able to appropriate part of it to solidify further their positions and concomitant authority and power. In good times, they can probably store enough surplus food to alleviate crises caused by drought and other adverse environmental conditions. In bad times, they have first claim on this surplus, and their fields will still be maintained by the villagers. This differential access to valued resources, like other attributes of Pueblo society discussed above, supports the hypothesis that the Pueblos are nonegalitarian and should be classified among the rank and stratified societies.

Authority and Power Among Pueblo Priests

The central issue in this section is whether Pueblo priests have only authority (as defined earlier) or both authority and power, whether or not they can "channel the behavior of others by the threat of use of sanctions" (Fried 1967:13). A review of the literature indicates that, contrary to McGuire's assertion, Pueblo priests do have both impressive authority and power.

Hoebel (1969:96) argues that the Keresan pueblos have centralized coercive power: "thought control and behavioral conformity [are] . . . managed by means of an insistent mechanism of public law, which in the Keresan pueblos is administered by a theocratic society" (Hoebel 1969:98). Ultimate power within the pueblo is vested in the "Sacerdotal Council" consisting of the medicine society heads and the cacique (Hoebel 1969:99). Those who do not conform to expected behavior may initially be subjected to gossip and ridicule; if their behavior continues they may receive a lecture by a priest (Hoebel 1969:102; cf. Dozier 1954:339-353; Parsons 1939:153). If these relatively mild sanctions do not bring the transgressor into conformity with behavioral rules, the person may be whipped or fined. Payment of fines is often made in land, horses, clothing, and food (Hoebel 1969:102-104, 111).

In a case at Santa Clara, a man who earned a living from nontraditional means (iron and woodworking) and who refused to take part in pueblo customs was severely punished. During a protracted period of confrontation, attempts were made to banish him from the pueblo; he eventually lost all his land and material possessions. His sons were also punished for his "sins" and eventually left the pueblo at their father's urging (Lange 1982:193-

194). Hill (in Lange 1982:194-196) reports other cases of punishment for failure to behave according to pueblo rules.

Ultimately, persons who refuse to do as they are expected and told can be charged with witchcraft: "To be charged with witchcraft was to be sentenced to death" (Hoebel 1969:110). Ethnographic accounts of persons charged with witchcraft are available for all the Pueblos, and the trials and executions of suspected witches are documented in the literature (Cushing 1883:42-43; Dozier 1954:339-341; Hoebel 1969:99, 110; Lange 1959:458-463, 1982:334-336; Parsons 1939:62-69; Simmons 1942; Simmons 1974; Stevenson 1904:393). Priests conduct the witchcraft trials and carry out the sentences, including execution of the wrongdoer (e.g., Cushing 1882a, 1883; Hoebel 1969).

The foregoing indicates that Pueblo priests have both authority and power. These data further support the hypothesis that the Pueblos are nonegalitarian.

Limits to Priestly Authority and Power

The authority and power of priests are not absolute; there are checks and balances at every level of the social hierarchy, and priests, like common people, are subject to this system.

> By the townspeople themselves the chiefs are far from being unnoticed. Close-mouthed as are the Pueblos toward strangers, among themselves they are prime gossips, and much of their gossip is over those who are the priests....
>
> Absenteeism on the part of any Town Chief would be talked about. . . .
>
> Incontinence during ceremonial is an even more scandalous offense than absenteeism. Crow-Wing refers to a former Powamu society chief as not a good chief. "He got into trouble with a woman." This meant that the people would be crippled with rheumatism. . . . Once a Zuni Pekwin "lasted" only half a year because of immoderation. . . . Chiefs are peculiarly open to the charge of witchcraft. "The best men of the pueblo may be witches" (Laguna). . . . Ts'awele himself, War chief and therefore lord high prosecutor of witches, for many years the most commanding personality in Zuni, was charged while conducting a witch case of keeping owl feathers inside his moccasins and was demoted from his high office. . . .
>
> Since the character of the chiefs should be above suspicion and a man's behavior affects the validity of his ceremony, town gossip, or much of it, is not idle or barren; it is public opinion bringing pressure upon those whose behavior or character is believed to be vital to the efficacy of the ritual and so of supreme interest to all the townspeople, to their health and to their crops [Parsons 1939:152-154].

It is also believed that priests who perform their duties inadequately can cause drought, windstorms, damaging hail, disease, and other misfortune (Parsons 1925a:75; Titiev 1938). As noted in the above quotation from Parsons, and elsewhere, such persons are likely to be removed from office.

> Several years ago the droughts were very serious, and a retired sun priest was suspected

> and impeached, and his place was filled by another. The people whispered among themselves, "He is a sorcerer" [Stevenson 1904:393].

If the offense is serious or frequent enough, or if the results of a priest's presumed misconduct are extreme, the guilty priest may be charged with witchcraft and executed by other priests. Among the Keres, Bandelier (in Hoebel 1969:99) states that the cacique could be impeached by the war priest and that a deposed cacique seldom lived long; his knowledge (power) could not be tolerated for fear of witchcraft (Parsons 1939:62, 154, 491). The war chief carried out the execution, and only the initiated were aware that the cacique's death was other than natural.

Thus, Pueblo priests are subject to limitations on their authority and power. These limitations are similar to those faced by the lay public, i.e., both priests and common people are expected to conform to the rules of acceptable behavior for their respective roles and to meet their respective responsibilities. In both cases, failure to behave as expected invokes a series of increasingly severe sanctions that, ultimately, can lead to the death of the wrongdoer.

Burial Practices

Archaeologists are keenly interested in burial practices as a key to understanding social differences, including rank and stratification (Brown 1971; Chapman et al. 1981; Peebles and Kus 1977; Whittlesey 1978). Generally they argue that the higher the rank of the individual, the more important the symbols, the more complex the burial, and the greater the expenditure of energy in the mortuary rites (Peebles and Kus 1977).

In this context, the Pueblos are atypical of nonegalitarian societies. As discussed previously (Reyman 1978:249-250; cf. Frisbie 1978), Pueblo dead are buried with relatively few grave goods, and relatively little energy is expended on anyone's funeral regardless of status (rank) when alive.

> Though the Hopi have a great fear of death and of dead people, mortuary rites are extremely simple in character. There is no mortuary feasting or disposal of food through gifts. The clothes in which the individual died are left on the body. Buckskin or wedding blanket is wrapped around the corpse and the body is quickly buried. Clothes and small items of personal property of the deceased are thrown away. The rest of the property owned by the individual is disposed of according to the patterns summarized earlier [Beaglehole 1937:77; cf. Simmons 1942].

The other Pueblos are similar in this regard. High rank, if indicated at all, is more likely to be indicated by special body paint (designs) than by grave goods (Goldfrank 1927:65-66). If it could be demonstrated that these designs are very important symbols not available to others, then perhaps this aspect of Pueblo mortuary ritual could be construed as conforming to Peebles and Kus's model. As Beaglehole and others (Parsons 1939:90; Stevenson 1904:291, 307) note, most possessions are given away or destroyed rather than buried with the dead.

Occasionally, elaborate burials do occur, although these are clearly exceptions in response to special circumstances. The appearance of a "death fly" at the grave of a Hopi child resulted in a relatively elaborate burial in terms of grave goods in an attempt to ward off death from other family members (Simmons 1942:288). At Zuni, the death of a transvestite named We'wha, one of the tallest and the strongest members of the pueblo and one who served as the chief personage in many ceremonies, was followed by an elaborate burial by Zuni standards: numerous wraps, a blanket, three silver necklaces, turquoise earrings, and many "bangles" (Stevenson 1904:310-313). This is hardly an elaborate burial, however, when compared with some of the prehistoric burials at Pueblo Bonito (Pepper 1909), Aztec Ruin (Morris 1924:193-195), and Ridge Ruin (McGregor 1943).

It is possible, even probable, that other elaborate, historical Pueblo burials of high-rank individuals have gone unreported. Nevertheless, a search of the ethnographic literature cited throughout this paper and also of other, unpublished sources indicates that, unlike many other rank and stratified societies, the Pueblos do not mark high rank by differential burial practices to any significant extent. Although important symbols (body paint) may be used to indicate rank, neither burial complexity nor energy expenditure is substantially greater, if at all, for priests than for lay members of the pueblo.

Given that very elaborate burials do occur prehistorically, indicating stratification in earlier times (Akins and Schelberg 1981; Reyman 1978), the lack of corresponding burials reported in the ethnographic literature is noteworthy. The complexity of burial practices, like other features of Pueblo culture, appears to have attenuated sharply since the Spanish entradas, an inference consistent with some earlier work (Reyman 1971, 1976b), and with one of Upham's (1982) major conclusions.

SOME MAJOR IMPLICATIONS FOR THE CHACOAN ANASAZI

This paper is focused mainly on the historical Pueblos. This study has some implications, however, for Chacoan archaeology. There is continuing discussion over whether the Chacoan pueblos, especially the large town sites, represent egalitarian or nonegalitarian (either rank or stratified) societies (Akins and Schelberg 1981; Ellis 1983:xxxv; Judge et al. 1981; Lekson 1981; Reyman 1971, 1980; Vivian 1970). Although the archaeological evidence is hardly unequivocal, and there is no absolute agreement among archaeologists as to its interpretation, there seems to be a tendency to consider such sites as Pueblo Bonito, Chetro Ketl, and Peñasco Blanco as the products of stratified societies; at the very least, there is a social hierarchy implicit in the size and layout of these large pueblos and in the apparent planning needed to construct them. Social hierarchy is also implicit in the distribution of materials within these sites (Akins and Schelberg 1981; Reyman 1971, 1978, 1980; cf. Judge et al. 1981; Lekson 1984; Vivian 1970). This inference seems reasonable, especially when the large sites are compared with the smaller pueblos (Bc-50, Bc-51, etc.) in the canyon. Whether this apparent rank or stratified society developed as a result of Mexican influences on the Chacoan Anasazi or primarily in response to relatively localized conditions, the conclusion is the same: the degree of cultural complexity evident at Chaco Canyon is best explained in terms of general and overall direction and control by some superordinate political authority and power.

In relation to the archaeoastronomy focus of this volume, given the probability of a preconceived plan for Pueblo Bonito, it is unlikely that the winter solstice alignments of the windows in Rooms 225B and 228B-SE could have been incorporated into the construction without the presence of an individual or group with the authority and power to order and direct the work (Reyman 1976b, 1980). Someone, presumably a priest, made the necessary solar observations, and then this person or some other person in authority directed the construction of the rooms and windows within the preplanned roomblocks to incorporate these alignments.

The rooms with solstice alignments occur in two sections of the pueblo that were built at different times. The expansion of Kiva C prevented continued use of the observation window in Room 228B-SE (Reyman 1976a); therefore, to avoid interruption in the observational system and the calendrically scheduled activities, the window in Room 225B had to be ready for use before Room 228B-SE could be partitioned and filled with debris to support the south wall of Kiva C. Coordination of such an activity strongly suggests centralized control. Similar coordination can be inferred from the other Chacoan astronomical alignments, such as that at Chetro Ketl, and in the relationship of Casa Rinconada to mesa-top cairns (Reyman 1971, 1984).

Although we do have ethnographic data for centralized direction of such communal projects as irrigation ditch maintenance and kiva construction and repair, I am not aware of any evidence pertaining to the construction of astronomically aligned buildings or architectural features. We have only the results of such behavior in the form of the "suntower" at Zuni (Lange and Riley 1970:69-70) and, formerly, the Flint Ku-sha′lī House at Cochiti (Lange 1959:55-56, Plate 8b). The nature of these structures and of astronomically aligned features at other pueblos indicates that directed communal labor was used to construct them, but we have no actual accounts describing the work. In this context, therefore, the ethnographic data are of limited value for generating hypotheses to be tested at archaeological sites such as those at Chaco Canyon.

I suggest that we not expend much more of our energy and limited financial resources doing fieldwork to demonstrate the existence, per se, of astronomical alignments at prehistoric sites. Like their historical descendents, the prehistoric Anasazi and others used astronomy to establish a calendar for organizing and executing subsistence and ceremonial activities. We can take this as given. Let us now proceed to the more important task of studying the implications of astronomical knowledge for social stratification, centralized control, cultural ecology, etc. We should put behind us the "shotgun" approach to searching for alignments (Reyman 1975, 1979) and the "I'll show you my alignment if you show me yours" syndrome that all too often characterizes much of the research on Native American astronomy.

The ethnographic data for Pueblo burial practices are similarly of limited value in helping us to understand Chacoan Anasazi mortuary behavior and its implications for social differentiation and ranking. Elaborate burials have been unearthed in the Southwest, some of them in Chaco Canyon. The burials in Rooms 32 and 33 at Pueblo Bonito include thousands of artifacts; they are clearly elaborate by New World standards and are especially elaborate relative to general Chacoan Anasazi mortuary practices (Reyman 1978; Reyman and Nickels 1984). They are burials one might expect to find in conjunction with a stratified society, particularly when they are considered within the context of the other archaeological data from the site and from related, contemporaneous sites within the canyon and its environs. Although some think that these burials do represent the remains of high-ranked individuals, especially the burials of the two males in the subfloor tomb in Room 33 (Pepper 1909; Reyman 1978; Reyman and Nickels 1984; cf. Akins and Schelberg 1981; Judge et al. 1981), the fact is that all fifteen burials in Rooms 32 and 33 are quite unlike any Pueblo burials recorded in the published ethnographic literature in terms of both the quantity and quality of the grave goods. These Pueblo Bonito interments fit well with the correlates proposed by Peebles and Kus (1977; cf. Akins and Schelberg 1981; Reyman 1978; Reyman and Nickels 1984). The differences are even more pronounced when we examine the burials of historical Pueblo priests. As noted earlier, the paucity of grave goods and the lack of mortuary rites associated with Pueblo priests are unusual in the context of a nonegalitarian society.

Thus, the ethnographic data are of limited use for the study of the prehistoric context at Chaco Canyon, except to indicate that the burials in Rooms 32 and 33 are atypical relative to known Pueblo funerary practices. Despite the well-documented prehistoric-historical continuum in the Southwest, there is marked discontinuity in burial practices between the Anasazi and their historical Pueblo descendents. We must not only develop alternative hypotheses to explain remarkable burials at Pueblo Bonito and elsewhere (e.g., at Aztec and Ridge ruins), but we must consider the problem of discontinuity in burial practices as well.

SUMMARY AND CONCLUSIONS

It has been demonstrated that the Pueblos are nonegalitarian societies characterized by centralized leadership perpetuated through descent and hereditary succession. The Pueblos meet the qualifications discussed by Fried for rank society and also, in some cases, for stratified societies. Priestly authority, power, and social stratification are backed by physical coercion and based on access to and control of valued resources (land, water, communal labor), as well as on esoteric ritual knowledge.

The historical Pueblos are atypical of nonegalitarian societies in that ranking and stratification are not reflected in their mortuary practices. This may be another example of the attenuation of earlier, more complex behavior as a result of acculturative pressures from Euroamerican society, but alternatively this attenuation may have begun considerably earlier, perhaps as early as the fourteenth century, as Upham argues.

This atypical feature has an unfortunate implication for Chacoan archaeology because it precludes the use of Puebloan ethnographic data for generating hypotheses to explain the elaborate burials at Pueblo Bonito (and elsewhere). On a positive note, by demonstrating that this break in the presumed prehistoric-historical continuum occurred, we should now be better able to study the development at Chaco Canyon as something more than a mere precursor of the later, historical Pueblos. The prehistoric situation found at Chaco Canyon probably represents a level of complexity and integration far beyond that recorded by anthropologists over the last century for the historical Pueblos, or even by the Spanish several hundred years earlier. It is toward understanding this prehistoric development and the later changes that we must continue to work.

ACKNOWLEDGMENTS

The research for this paper was supported by NSF grants GS-2829, GS-40410, and BNS 77-08070; by Wenner-Gren for Anthropological Research grant-in-aid 4102; and by Illinois State University. The comments of the symposium participants are gratefully acknowledged, as are those of Charles H. Lange and especially C. Randall Morrison. The author alone is responsible for any and all errors of fact or interpretation.

REFERENCES CITED

Akins, Nancy J., and John D. Schelberg
1981 Evidence for Organizational Complexity as Seen from the Mortuary Practices of Chaco Canyon. Paper presented at the 46th Annual Meeting of the Society for American Archaeology, San Diego.

Beaglehole, Ernest
1937 Notes on Hopi Economic Life. Yale University Publications in Anthropology No. 15. New Haven.

Benedict, Ruth
1934 Patterns of Culture. Houghton-Mifflin, Boston.
1935 Zuñi Mythology, 2 vols. Columbia University Contributions to Anthropology 21. New York.

Bradfield, Maitland
1971 The Changing Pattern of Hopi Agriculture. Royal Anthropological Institute Occasional Paper No. 30. London.

Brandt, Elizabeth A.
1980 On Secrecy and Control of Knowledge: Taos Pueblo. In Secrecy: A Cross-Cultural Perspective, edited by Stanton K. Tefft, pp. 123-146. Human Sciences Press, New York.

Brown, James A., editor
1971 Approaches to the Social Dimensions of Mortuary Practices. Society for American Archaeology Memoir 25. Washington, D. C.

Bunzel, Ruth L.
1932 Introduction to Zuni Ceremonialism. Forty-Seventh Annual Report of the Bureau of American Ethnology, pp. 467-544. Smithsonian Institution, Washington, D. C.

Chapman, Robert, Ian Kinnes, and Klavs Randsborg, editors
1981 The Archaeology of Death. Cambridge University Press, Cambridge.

Colton, Harold S.
1934 A Brief Survey of Hopi Common Law. Museum of Northern Arizona Museum Notes 7(6):21-24. Flagstaff.

Colton, Mary-Russell F., editor
1936 Truth of a Hopi, by Edmund Nequatewa. Museum of Northern Arizona Bulletin No. 8. Flagstaff.

Cushing, Frank H.
1882a My Adventures in Zuñi (Part I). The Century Illustrated Monthly Magazine 25:191-207.
1882b My Adventures in Zuñi (Part II). The Century Illustrated Monthly Magazine 25:500-511.
1883 My Adventures in Zuñi (Part III). The Century Illustrated Monthly Magazine 26:28-47.

Dozier, Edward P.
1954 The Hopi-Tewa of Arizona. University of California Publications in American Archaeology and Ethnology 44(3). Berkeley and Los Angeles.
1970 The Pueblo Indians of North America. Holt, Rinehart and Winston, New York.

Eggan, Fred
1950 Social Organization in the Western Pueblos. University of Chicago Press, Chicago.

Ellis, Florence Hawley
1983 Foreword. In The Architecture and Dendrochronology of Chetro Ketl, Chaco Canyon, New Mexico, edited by Stephen H. Lekson, pp. xxiii-xxxviii. Reports of the Chaco Center No. 6. Division of Cultural Research, National Park Service, Albuquerque.

Ellis, Florence Hawley, and Laurens Hammack
1968 The Inner Sanctum of Feather Cave, a Mogollon Sun and Earth Shrine Linking Mexico and the Southwest. American Antiquity 33:25-43.

Forde, C. Daryll
1931 Hopi Agriculture and Land Ownership. Journal of the Royal Anthropological Institute 61:357-405.
1963 Habitat, Economy, and Society. E. P. Dutton, New York.

Fried, Morton H.
1967 The Evolution of Political Society: An Essay in Political Anthropology. Random House, New York.

Frisbie, Theodore R.
1978 High Status Burials in the Greater Southwest: An Interpretive Synthesis. In Across the Chichimec Sea: Papers in Honor of J. Charles Kelley, edited by Carroll L. Riley and Basil C. Hedrick, pp. 202-226. Southern Illinois University Press, Carbondale.

Glenn, James R.
1982 Register to the Papers of Neil Merton Judd. Smithsonian Institution, Washington, D. C.

Goldfrank, Esther S.
1927 The Social and Ceremonial Organization of Cochití. American Anthropological Association Memoir No. 33. Menasha, Wisconsin.

Hieb, Louis A.
1979 The Ritual Clown: Humor and Ethics. In Forms of Play of Native North Americans, edited by Edward Norbeck and Claire R. Farrer, pp. 171-188. 1977 Proceedings of the American Ethnological Society. West, St. Paul.

Hoebel, E. Adamson
1969 Keresan Pueblo Law. In Law in Culture and Society, edited by Laura Nader, pp. 92-116. Aldine, Chicago.

Judge, W. James, H. Wolcott Toll, William B. Gillespie, and Stephen H. Lekson
1981 Tenth Century Developments in Chaco Canyon. In Collected Papers in Honor of Erik Kellerman Reed, edited by Albert H. Schroeder, pp. 65-91. Papers of the Archaeological Society of New Mexico No. 6. Albuquerque.

Kessell, John L.
1979 Kiva, Cross, and Crown: The Pecos Indians and New Mexico, 1540-1840. National Park Service, Washington, D.C.

Kroeber, Alfred L.
1917 Zuñi Kin and Clan. Anthropological Papers of the American Museum of Natural History 17(2). New York.

Ladd, Edmund J.
1979 Zuni Social and Political Organization. In Southwest, edited by Alfonso Ortiz, pp. 482-491. Handbook of North American Indians, Vol. 9. Smithsonian Institution, Washington, D.C.

Lange, Charles H.
1959 Cochití: A New Mexico Pueblo Past and Present. University of New Mexico Press, Albuquerque.
1982 An Ethnography of Santa Clara Pueblo, New Mexico, by W. W. Hill, edited and annotated by Charles H. Lange. University of New Mexico Press, Albuquerque.

Lange, Charles H., and Carroll L. Riley
1970 The Southwestern Journals of Adolph F. Bandelier: 1883-1884. University of New Mexico Press, Albuquerque.

Lekson, Stephen H.
1981 Standing Architecture and the Interpretation of Local and Regional Organization of Chaco Canyon. Paper presented at the 46th Annual Meeting of the Society for American Archaeology, San Diego.
1984 Great Pueblo Architecture of Chaco Canyon. Publications in Archeology 18B, Chaco Canyon Studies. National Park Service, Santa Fe.

McCluskey, Stephen C.
1977 The Astronomy of the Hopi Indians. Journal for the History of Astronomy 8(2):174-195.
1981 Transformations of the Hopi Calendar. In Archaeoastronomy in the Americas, edited by R. A. Williamson, pp. 173-182. Ballena Press. Los Altos, California.
1982 Historical Archaeoastronomy: The Hopi Example. In Archaeoastronomy in the New World, edited by A. F. Aveni, pp. 31-57. Cambridge University Press, Cambridge.

McGregor, John C.
1943 The Burial of an Early American Magician. Proceedings of the American Philosophical Society 86(2):270-298.

McGuire, Randall H.
1983 Review of Polities and Power: An Economic and Political History of the Western Pueblo. American Antiquity 48:651-652.

Mindeleff, Victor
1891 A Study of Pueblo Architecture: Tusayan and Cibola. In Eighth Annual Report of the Bureau of American Ethnology, pp. 3-228. Smithsonian Institution, Washington, D. C.

Morris, Earl H.
1924 Burials in the Aztec Ruin. Anthropological Papers of the American Museum of Natural History 26(3). New York.

Ortiz, Alfonso
1969 The Tewa World: Space, Time, Being, and Becoming in a Pueblo Society. University of Chicago Press, Chicago.

Parsons, Elsie Clews
1925a A Pueblo Indian Journal, 1920-1921. American Anthropological Association Memoir No. 32. Menasha, Wisconsin.
1925b The Pueblo of Jemez. Papers of the Southwestern Expedition No. 3. Phillips Academy, Andover, Massachusetts.
1929 The Social Organization of the Tewa of New Mexico. American Anthropological Association Memoir No. 36. Menasha, Wisconsin.
1932 Isleta, New Mexico. In Forty-Seventh Annual Report of the Bureau of American Ethnology, pp. 193-466. Smithsonian Institution, Washington, D. C.
1939 Pueblo Indian Religion, 2 vols. University of Chicago Press, Chicago.

Parsons, Elsie Clews, editor
1936 Hopi Journal of Alexander M. Stephen, 2 vols. Columbia University Contributions to Anthropology Vol. 23. New York.

Peebles, Christopher S., and Susan M. Kus
1977 Some Archaeological Correlates of Ranked Societies. American Antiquity 42:421-448.

Pepper, George H.
1909 The Exploration of a Burial Room in Pueblo Bonito, New Mexico. In Putnam Anniversary Album, pp. 196-252. G. E. Stechert, New York.

Reyman, Jonathan E.
1971 Mexican Influence on Southwestern Ceremonialism. Ph.D. dissertation, Southern Illinois University, Carbondale. University Microfilms, Ann Arbor.
1975 The Nature and Nurture of Archaeoastronomical Studies. In Archaeoastronomy in Pre-Columbian America, edited by Anthony F. Aveni, pp. 205-215. University of Texas Press, Austin.
1976a Astronomy, Architecture, and Adaptation at Pueblo Bonito. Science 193:957-962.

1976b Mexican-Southwestern Interaction: The Puebloan Ethnographic Evidence. In Archaeological Frontiers: Papers on New World High Civilizations in Honor of J. Charles Kelley, edited by Robert B. Pickering, pp. 87-128. University Museum Studies No. 4. Southern Illinois University Museum, Carbondale.

Reyman, Jonathan E.
1978 Pochteca Burials at Anasazi Sites? In Across the Chichimec Sea: Papers in Honor of J. Charles Kelley, edited by Carroll L. Riley and Basil C. Hedrick, pp. 242-259, 273 (notes 14-16). Southern Illinois University Press, Carbondale.
1979 Some Observations on Archaeology and Archaeoastronomy. Archaeoastronomy 2(2):11-13.
1980 The Predictive Dimension of Priestly Power, with a Note on Solar and Lunar Cycles by Harold J. Born. In New Frontiers in the Archaeology and Ethnohistory of the Greater Southwest, edited by Carroll L. Riley and Basil C. Hedrick, pp. 40-59. Transactions of the Illinois State Academy of Science 72(4). Illinois State Museum, Springfield.
1982 Adaptation, Astronomy, and Architecture at Mesa Verde. Paper presented at the Forty-Seventh Annual Meeting of the Society for American Archaeology, Minneapolis.
1984 Anasazi Adaptation. Ms. in author's possession, Illinois State University, Normal, Illinois.

Reyman, Jonathan E., and Martin K. Nickels
1984 The Burials of Pueblo Bonito. Ms. in authors' possession, Illinois State University, Normal, Illinois.

Saquet, Janette
1983 Register to the Papers of Frank Harold Hanna Roberts, Jr. Smithsonian Institution, Washington, D.C.

Simmons, Leo W., editor
1942 Sun Chief: The Autobiography of a Hopi Indian. Yale University Press, New Haven.

Simmons, Marc
1974 Witchcraft in the Southwest: Spanish and Indian Supernaturalism on the Rio Grande. University of Nebraska Press, Lincoln.

Smith, M. Estellie
1969 Governing at Taos Pueblo. Eastern New Mexico University Contributions to Anthropology 2(1). Portales.

Sofaer, A., R. M. Sinclair, and L. E. Doggett
1982 Lunar Markings on Fajada Butte, Chaco Canyon, New Mexico. In Archaeoastronomy in the New World, edited by A. F. Aveni, pp. 169-181. Cambridge University Press, Cambridge.

Stevenson, Matilda Coxe
1904 The Zuñi Indians: Their Mythology, Esoteric Fraternities, and Ceremonies. Twenty-Third Annual Report of the Bureau of American Ethnology. Smithsonian Institution, Washington, D. C.

Titiev, Mischa

1938 Dates of Planting at the Hopi Indian Pueblo of Oraibi. Museum of Northern Arizona Museum Notes 11(5):39-42.

1944 Old Oraibi: A Study of the Hopi Indians of Third Mesa. Peabody Museum of American Archaeology and Ethnology Papers Vol. 22. Harvard University, Cambridge.

1972 The Hopi Indians of Old Oraibi: Change and Continuity. University of Michigan Press, Ann Arbor.

Udall, Louise

1969 Me and Mine: The Life Story of Helen Sekaquaptewa as Told to Louise Udall. University of Arizona Press, Tucson.

Upham, Steadman

1982 Polities and Power: An Economic and Political History of the Western Pueblo. Academic Press, New York.

Vivian, R. Gwinn

1970 An Inquiry into Prehistoric Social Organization in Chaco Canyon, New Mexico. In Reconstructing Prehistoric Pueblo Societies, edited by William A. Longacre, pp. 59-83. School of American Research, Santa Fe, and University of New Mexico Press, Albuquerque.

White, Leslie A.

1932a The Acoma Indians. Forty-Seventh Annual Report of the Bureau of American Ethnology, pp. 17-192. Smithsonian Institution, Washington, D. C.

1932b The Pueblo of San Felipe. American Anthropological Association Memoir No. 38. Menasha, Wisconsin.

1942 The Pueblo of Santa Ana, New Mexico. American Anthropological Association Memoir No. 60. Menasha, Wisconsin.

1962 The Pueblo of Sia, New Mexico. Bureau of American Ethnology Bulletin 184. Smithsonian Institution, Washington, D. C.

Whittlesey, Stephanie M.

1978 Status and Death at Grasshopper Pueblo: Experiments Toward an Archaeological Theory of Correlates. Ph.D. dissertation, University of Arizona, Tucson. University Microfilms, Ann Arbor.

Chapter 9

The Evolution of Hohokam Ceremonial Systems

David R. Wilcox

Striking similarities are readily apparent in the basic structure of Mesoamerican and Southwestern cosmological systems. As sketched by Cohodas (1974), the metastructure of Mesoamerican cosmology consisted of a four-directional symbolism ordered in time by the solstices and equinoxes and marked by the cyclic counterclockwise movement of the sun and other heavenly bodies in perpetual daily and annual processes up into the upperworld and down into the underworld. South was associated with day, the winter solstice, and the upperworld; west was associated with sunset, the vernal equinox, and the surface of the earth; north was associated with night, the summer solstice, and the underworld; and east was associated with sunrise, the autumnal equinox, and the surface of the earth (Cohodas 1974:105).

Vogt (1981) has shown that mountains, holes in the ground, mountain passes, rocks, and trees served as points of reference in a Mayan sacred landscape connecting the upper and lower worlds with the surface of the earth. Shrines at these places served both as a means of communication with supernaturals and as boundary markers among distinct groups at several scales of social organization. Ortiz (1969) has described the "structure of structures in Tewa life" in very similar terms. Titiev's (1944) delineation of "those essential features which serve as common denominators in Hopi world view" defines a nearly identical model.

The ritual cycles of both the Pueblos and the Mesoamericans were calendrically structured (Titiev 1960), although the calendar in Mesoamerica was certainly far more complex and required much greater sophistication in the practice of astronomy (Aveni 1982). Yet it is the fact of calendrical ceremonialism that distinguishes Pueblo religion most clearly from those of other cultures in western North America.

My purpose in this paper is to show that the prehistoric Hohokam also seem to have developed much the same metastructure of cosmological belief as the Pueblo and Mesoamerican cultures. Three classes of data reveal this most clearly: (a) the architecture of the Big House at Casa Grande ruins, (b) the changing site structure of Snaketown and certain other sites, and (c) the changing structure of the Hohokam regional system as delineated by the data on Hohokam ballcourts. Based on these data a model of the evolution of Hohokam ceremonial systems is constructed. This model suggests that a calendrical ceremonial system was created by the Hohokam at least by the time of the Sedentary period, and that Casa Grande was quite possibly an astronomical observatory. These findings support the hypothesis of Southwestern participation in the Mesoamerican tradition.

CUSHING AND CASA GRANDE

Cushing was the first anthropologist to feel the resonance between the world view of a modern Southwestern society, the Zuni, and the relational structure of Hohokam settlement data. In 1887, after five years at Zuni (Green 1979), Cushing possessed a greater and more intimate knowledge of Pueblo ritual belief than any subsequent Hohokam archaeologist. After only a few weeks of exploring Hohokam sites in the Salt River Valley, Cushing was on the platform mound at La Ciudad de los Hornos when across the valley he noticed several other mounds.

> These . . . struck me suddenly as possibly pertaining to a system like the system of pueblos which originally composed the "Seven Cities of Cibola" [at Zuni]. . . . It occurred to me that if my sudden inference were true the shrines of sacrifice to the Six Regions--North, West, South, East, Upper, and Lower--would be distributed around each one of these ancient cities, as in Zuni according to the cardinal points symbolized, and the intermediate points for the Upper and Lower [Cushing 1892a: 109-110; also see Ortiz 1969; Vogt 1981].

It was then that he noticed Double Butte, where he soon found a cave containing hundreds of cane cigarettes; pieces of wooden birds similar to those later found in Chetro Ketl, Chaco Canyon (Vivian et al. 1978); and a wooden stick with a masked figure painted on it (Haury 1945). Cushing (1892a:115) felt that the artifact assemblage in this shrine was very reminiscent of Zuni winter solstice rites.

When he visited the ruins in December 1887, Cushing recognized in the floor plan of Big House (Figure 9.1) a relational structure very like the pattern created during the consecration of a Pueblo cornfield (Figure 9.2). He hypothesized that a similar structure was symbolized at Casa Grande (Cushing 1892b).

More recently the special character of Casa Grande has stimulated another hypothesis consistent with Cushing's. Molloy (1969) has suggested that Casa Grande was an astronomical observatory. Peculiar holes built into the walls seemingly served to permit precise observations of summer solstices, equinoxes, and lunar events. While a greater precision of measurement is required to test Molloy's conclusions (John Evans, personal communication 1981), my own analysis of the architecture of Casa Grande (Wilcox and Shenk 1977; Wilcox and Sternberg 1981) supports the plausibility of Molloy's conclusions.

The four-story structure represents a single building episode, and the wall openings studied by Molloy were original design features. Cushing's (1892b) model is two dimensional. It is possible, however, to redefine it using three dimensions, with the central tier of room spaces fulfilling the roles of nadir, middle, and zenith houses, respectively, and with the third-story rooms acting as the houses of the four cardinal directions. Interestingly, it is in the proposed "zenith" room that 10 of Molloy's observation holes are found, while the others are in the third story.

Figure 9.1. Plan of Casa Grande (after Cushing 1892b:309; Wilcox and Shenk 1977:28)

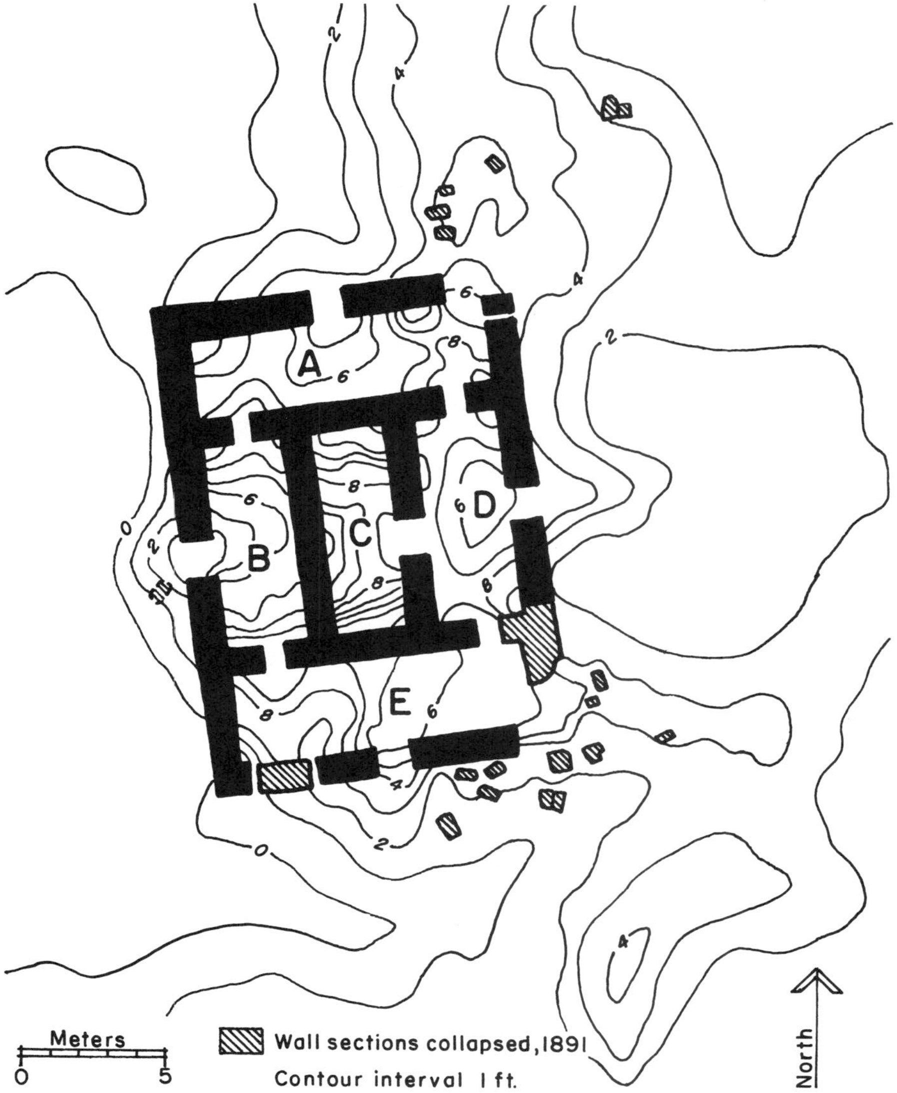

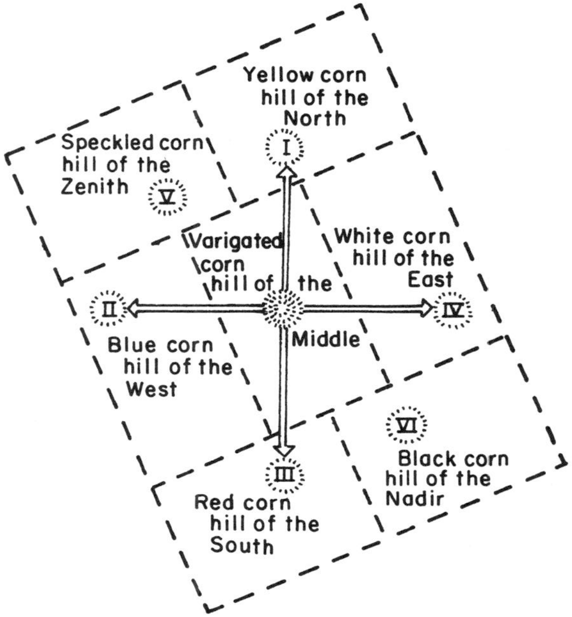

Figure 9.2. Diagram of a consecrated Hopi cornfield (after Cushing 1892b:304; Wilcox and Shenk 1977:26)

Just as we might predict from the metastructure model proposed by Cohodas (1974), the north room has the hole from which the summer solstice can be seen. Towerlike structures similar to the one at Casa Grande have recently been identified at La Ciudad (Wilcox 1986a) and Pueblo Grande (Cushing 1892a; Wilcox 1986b), and it seems likely that other sites also had them.

SNAKETOWN'S SITE STRUCTURE AND HOHOKAM SOCIAL ORGANIZATION

Casa Grande is one of the latest of the Hohokam structures, having been built during the late Classic period, sometime after AD 1300 (Wilcox and Shenk 1977). To understand its significance in the evolution of Hohokam ceremonialism it is necessary to examine an earlier site, that of Snaketown (Figure 9.3), which was occupied for a millenium or so from the beginning of the Pioneer period (which I place at about AD 200 ± 100 [see Cable and Doyel 1986; Wilcox and Shenk 1977] through the Colonial (AD 650-900) and Sedentary (AD 900-1100) periods. A case can be made that Snaketown served as a preeminent regional center throughout that time and was abandoned about AD 1100 at a time of major changes in the regional system (Haury 1976; Wilcox and Sternberg 1983).

Previous studies have suggested that Snaketown and other pre-Classic Hohokam sites were randomly structured. A systematic reexamination of the data on site structure, however, has shown that remarkable patterns are discernible (Wilcox et al. 1981). At Snaketown (Figure 9.3), one of the half-meter contour intervals on the topographic plan of the site defines a large rectangular area in the center of the village. Extensive excavations in this area showed that it contained no features and virtually no artifacts, and it has been identified as a central plaza. The rectangular outline of this plaza may have been imposed during the Sedentary period, when many of the large mounds that surround it were built (Haury 1976), but the general plaza area was already well defined by the time that the earliest houses known at Snaketown were built. Thus the plaza appears to have been a feature of the site throughout a millenium or so of occupation. The relations between other features and this plaza provide the data for a model of Snaketown's changing site structure.

Snaketown in the Early Pioneer Period

Three of the four earliest houses found at Snaketown are arrayed on three sides of the central plaza, with their doorways pointing toward it. It is possible that a fourth house on the east side was missed by the excavators. Four-direction symbolism like that found on the earliest Hohokam painted pottery may be indicated by these houses. Their unique size (the largest in the Hohokam sample) suggests that they may have had ceremonial significance on a regional scale, but the superposition of later Pioneer period houses on two of them shows that this relationship soon changed. What may be present, then, is a regional organization of social groups in different areas playing differential roles in the social system at the very time that pottery, clay figurines, and perhaps the ball game were introduced to the area from northern Mesoamerica (Wilcox 1986c; Wilcox and Sternberg 1983; cf. Cable and Doyel 1986).

Snaketown in the Late Pioneer Period

Irrigation was adopted in the Phoenix Basin during the late Pioneer period (Cable and Doyel 1985, 1986; Wilcox and Shenk 1977:

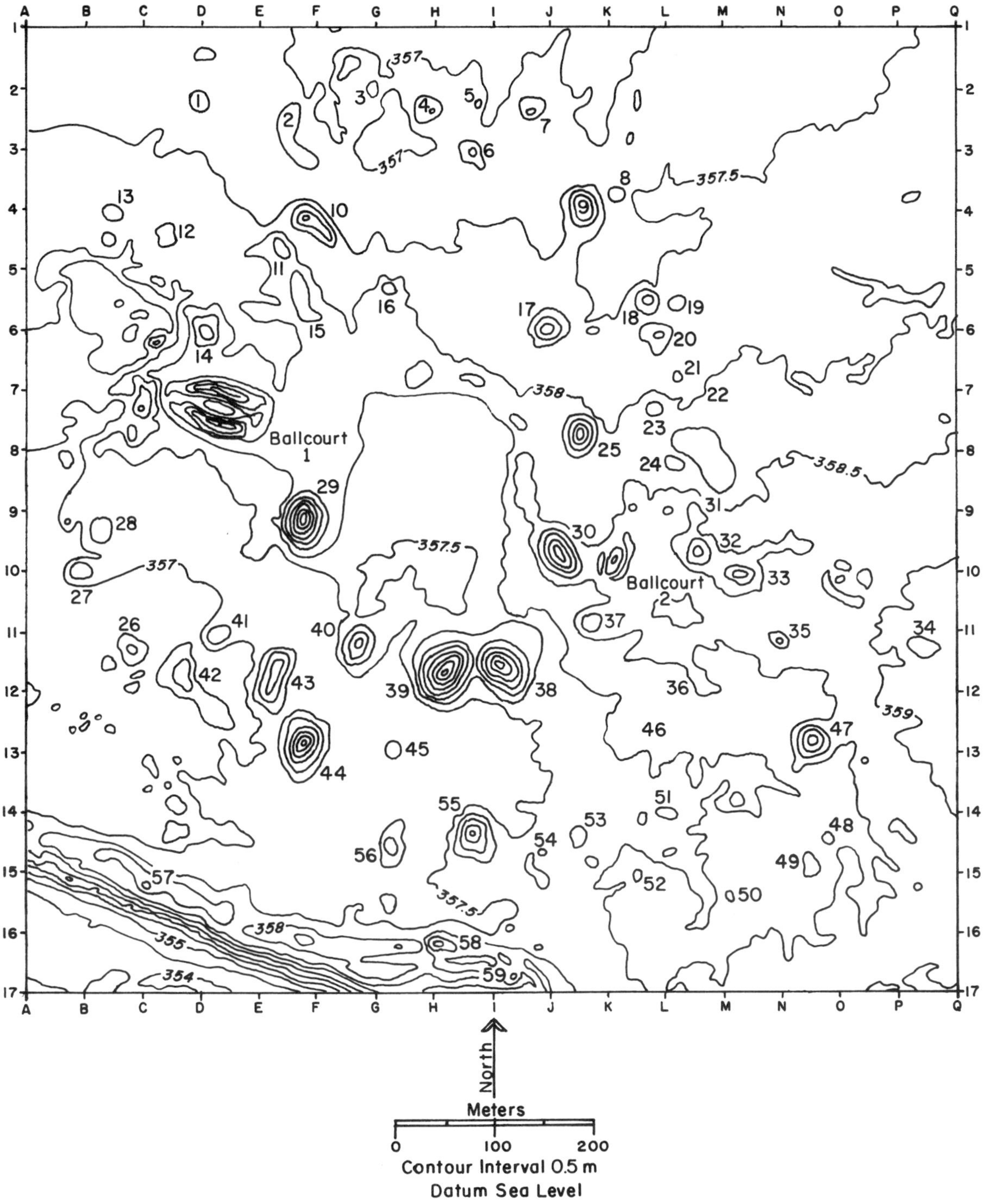

Figure 9.3. Map of Snaketown (after J. C. Fischer Motz's map in Gladwin et al. 1937:8; Wilcox et al. 1981:136)

180-182), and a series of communities that would endure for centuries as fixed loci of human activity were founded. The Pioneer period was the time of Hohokam figurines (Haury 1976:256). The appearance of figurines in the archaeological record may indicate that ancestors, and thus households as distinct units of production and consumption, acquired increased importance following the shift to sedentism and an agricultural strategy. Construction of a caliche-capped mound (Mound 40) toward the end of the Pioneer period at Snaketown, along with the presence of rare, esoteric ritual items (Gladwin et al. 1937:Plates LXXIX and LXXX), implies that specialists had begun to emerge and would eventually take over the mediation with supernaturals.

Snaketown in the Early Colonial Period

Fundamental changes occurred during the early Colonial period, a time of rapid population increase in the Phoenix Basin. The figurines that were so common during the Pioneer period disappear from the archaeological record just as the first ballcourts appear. Irrigation technology, which had permitted the population of the Phoenix Basin to grow disproportionately faster than that in neighboring areas, was adopted in some of the other areas, permitting their growth rates to equal that of the basin. Changes in the structure of regionwide mating networks seem likely. One indication of this might be a pattern observed at Snaketown, where valuables exotic to the Phoenix Basin that might have been used as bridewealth--items such as shell and argillite--were apparently deposited at a lower rate (relative to population size) during this period than they had been earlier (Wilcox and Sternberg 1983:230). Organizational change may have been a response to this reduction of access to exotic valuables in the Phoenix Basin. A new structure of authority on several scales of organization, with a reallocation of social advantage within the society, seems likely (see Chamberlin 1970). By the middle Colonial period, a special activity surface with a row of pits possibly related to the production of saguaro wine was built at Snaketown (Haury 1976). A new village-wide ceremony is indicated. Together with the artificial mounds and the large ballcourt, these data are evidence for an increasingly complex ceremonial system.

Snaketown in the Late Colonial Period

Population increase at Snaketown and in the Phoenix Basin seems to have leveled off during the late Colonial period, but significant new patterns of social organization continued to emerge. Cremation ceremonialism was transformed into a major institution. Stone palettes, stone censors, carved shells, and other ritual paraphernalia of these death rites were quickly adopted throughout the Phoenix Basin and adjacent areas. At the same time a great deal of exotic material was entering the basin. The new ritual practices significantly increased the distinctiveness of Hohokam culture and sharpened its cultural boundaries with other groups. The concomitant efflorescence of figurine use in the Prescott area (Morss 1954) further heightened the differences of cultural identity among what might be regarded as tribal groups.

Late in the Colonial period a special crematory floor, which provided a village focus for death rites, was superimposed on the earlier "saguaro wine" floor at Snaketown (Haury 1976:170-171). This cremation area was associated with a caliche-capped mound. The cremated bones were interred not just in one cemetery, however, but in at least five or six (Haury 1976:165). These facts suggest the emergence of still more formal social arrangements among intra-community groups in villages of at most only a couple of hundred people (Anderson 1986; Wilcox and Sternberg 1983:232-235). The fact that one of the cemeteries is much larger than the others (Haury 1976:165) may mean that one of the groups had a certain preeminence or dominance over the others. As yet, however, there is little other evidence to support any marked degree of social inequality.

Snaketown in the Sedentary Period

Population in the Phoenix Basin again increased rapidly during the Sedentary period, but there are indications that the net increase favored the Salt Valley over the Gila Valley (Wilcox and Sternberg 1983:201-202). Snaketown always had a concentric site structure, but now the core zone became more distinct from the remainder of the site with the construction of the central ring of mounds, the much higher percentage of large houses within that ring, and the greater complexity of their arrangements into house clusters (Figure 9.4; also see Wilcox et al. 1981). The boundedness of the village also became more distinct than formerly. Now the large, high, east-west ballcourt on the west; the small, low, north-south ballcourt on the east; the pair of large, high, caliche-capped mounds on the south (Mounds 38 and 39); and the small, low platform mound (Mound 16) on the north defined a series of dyadic relations that probably had great symbolic significance to the Hohokam: north/south, east/west, high/low, large/small, inner/outer. A functional differentiation of ceremonial facilities standing in a symbolically meaningful arrangement with respect to one another is indicated. Furthermore, Mounds 38 and 39 were built on top of the earlier crematory floor. A redefinition of the significance of death and the role of ancestors may be implied.

A division of Snaketown into northern and southern halves is also present in the Sedentary period site structure. Two extra-large houses, whose posthole patterns suggest the presence of benches, face one another on either side of a long, open area that may have been the end of a ceremonial racetrack (Figure 9.4). The core zone is thus divided in half. That these houses were special meeting houses or places of ritual is indicated by their size and by the discovery in the southern one of mountain sheep horn cores (Haury 1976:62). Given the many other dyadic relations already detailed, it does not seem entirely unwarranted to postulate a moiety organization (compare to Ortiz 1969).

The multiplicity of functionally differentiated ritual facilities implies that several ritual sodalities emerged during the Sedentary period, by around AD 1000. This would have radically redefined the structure of authority in Hohokam society and reallocated the distribution of social advantage. The distinctness of core-zone house clusters (Wilcox et al. 1981:169, 174) implies the emergence of an elite class that may have enjoyed unequal access to positions of authority and thus to the use of symbols of that authority. If we assume that storerooms for the ceremonial paraphernalia of the sodalities existed, it is clear that the chance destruction of one or more of these rooms might produce unique concentrations of exotic items in the archaeological record. In fact, House 8:6G (1934) at Snaketown may have been such a storeroom. It was built using exotic wood (Haury 1976:326) and contained all 28 of the copper bells found at Snaketown

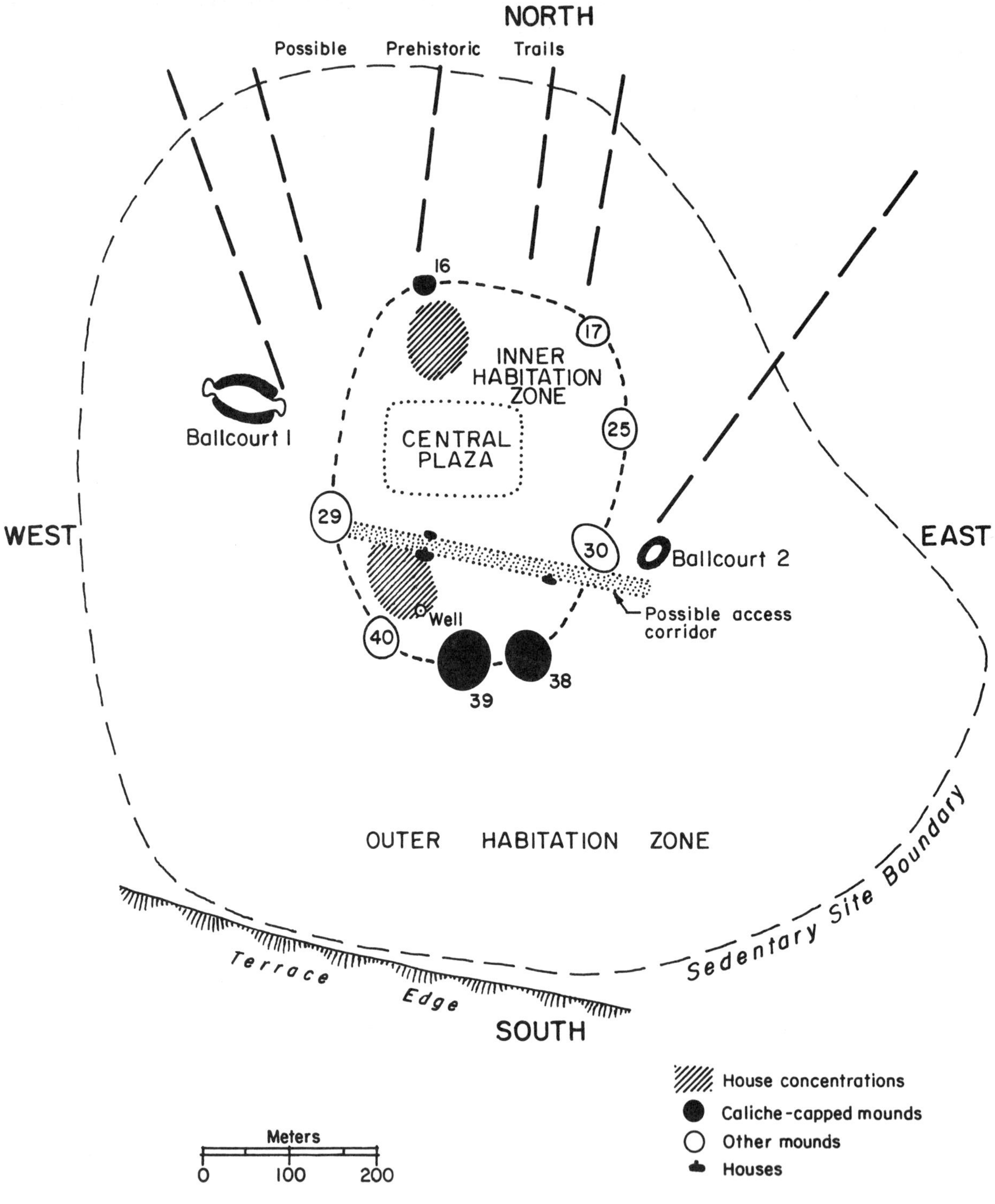

Figure 9.4. Plan of Snaketown during the Sedentary period (after Wilcox et al. 1981:142; Wilcox and Sternberg 1983:91)

as well as a considerable amount of other exotic goods (Nelson 1981).

Comparison of Site Structure at Snaketown with Other Hohokam Sites

The extent to which the patterns identified in Snaketown's site structure are replicated or complemented in other Hohokam sites is a major research problem now under intensive study as part of several large projects. Data currently available from a series of Hohokam sites within and outside the Phoenix Basin suggest that many did have site structures analogous to that of Snaketown (Wilcox 1985, 1986a; Wilcox and Sternberg 1983). Snaketown remains the most complex site yet described, however, and exact replication has not been found. Rather, similar elements are transformed, as in the process Levi-Strauss (1966) calls bricolage, to form a related series of structures.

BALLCOURTS AND THEIR DISTRIBUTION

One of the most distinctive forms of public architecture on Hohokam sites is an elliptical mound thought to be a ballcourt, a place where a version of the Mesoamerican ball game was played (Wilcox and Sternberg 1983). Haury (in Gladwin et al. 1937) first identified these mounds as ballcourts on the basis of excavation of the large court at Snaketown. It is the largest court ever built in the Hohokam area (Wilcox and Sternberg 1983:152).

More than 200 courts have now been recorded at more than 160 sites (Wilcox 1985). The first Hohokam courts were built in the early Colonial period and were widely distributed, occurring within a radius of 100 km or more in all four cardinal directions from Snaketown. By the time of the Sedentary period, a core/periphery structure in the regional social system had become well defined. Forty percent of the ballcourt sites were in the Phoenix Basin. About 30 sites had two or more courts that may have been contemporaneous, like the two at Snaketown (Wilcox et al. 1981:138-139). A study of mutual access among the ballcourt sites (Wilcox and Sternberg 1983:198-203) leads to the definition of four distinct zones (Figure 9.5): (a) a core consisting of the riverine communities within the Phoenix Basin and along the lower Verde River; (b) an inner periphery or rural hinterland of nonriverine sites within the Phoenix Basin and Buttes Dam areas, none of which have more than one court on the same site; (c) an intermediate periphery of separate local systems in the Gila Bend, middle Verde, lower San Pedro, and Tucson Basin areas; and (d) a far periphery of more distant populations, who did not share the Hohokam cultural identity but participated at different times and apparently in different ways in the Hohokam regional system. Hohokam courts are notably absent from the Papagueria and the Tonto Basin. The courts in the Flagstaff area may all date to a post-1100 period (Wilcox 1986d).

The distribution of Hohokam court sites in or near the Phoenix Basin in relation to plant communities (Figure 9.6) reveals many relationships that constitute structural conditions affecting Hohokam social life. Notable first is the striking closeness of adjacent sites; most are only about 5 or 6 km (3 mi) apart, although some are as far as 15-20 km apart. This closeness of spacing probably indicates a continuous social integration throughout the ballcourt network. At the end of the Sedentary period the great continuity in these linear arrays of Hohokam villages was severed when the lower Verde and Buttes Dam areas were abandoned and populations in the Gila Bend area shifted to new locations. In the Sedentary period,

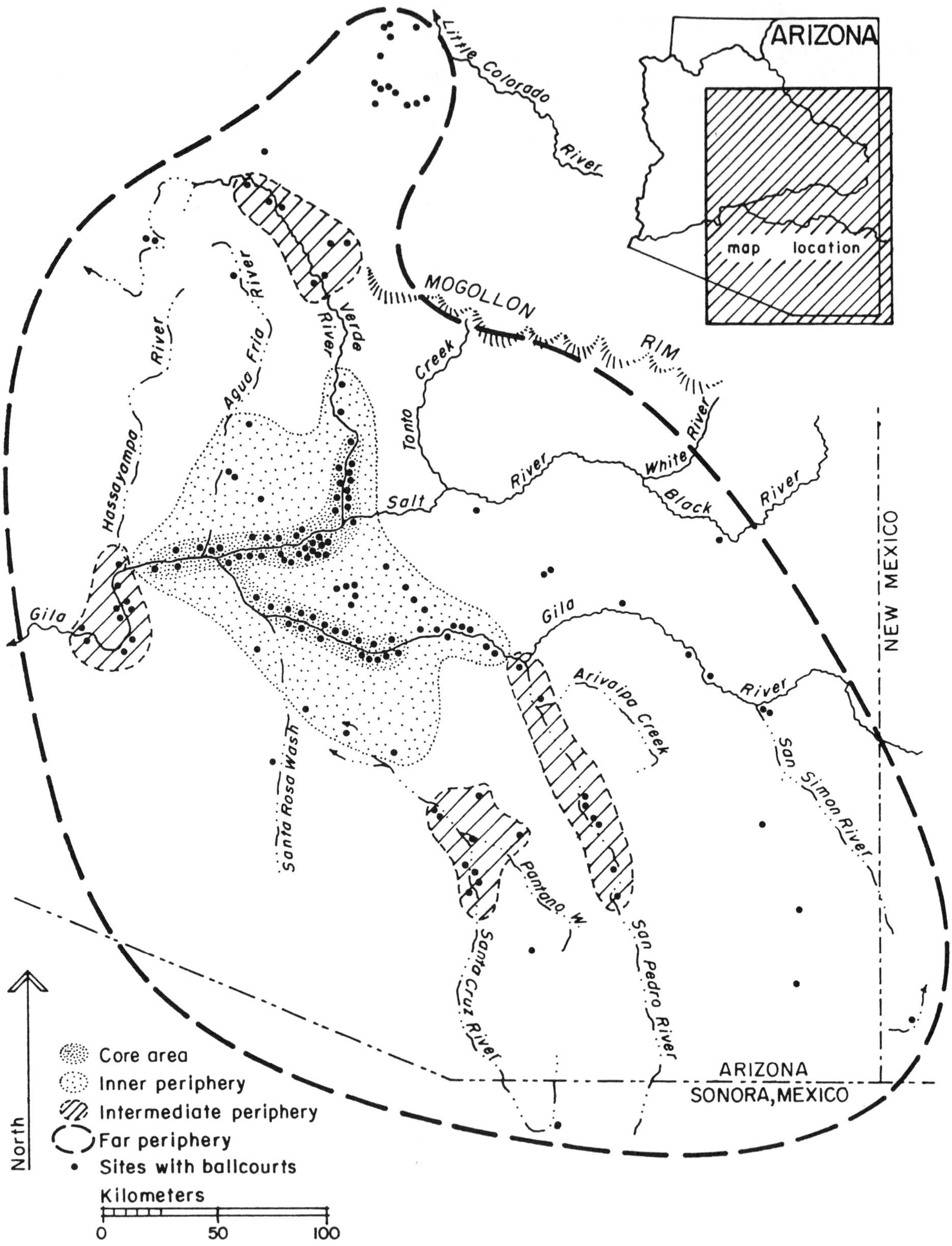

Figure 9.5. The Hohokam regional system during the Sedentary period (after Wilcox 1984b)

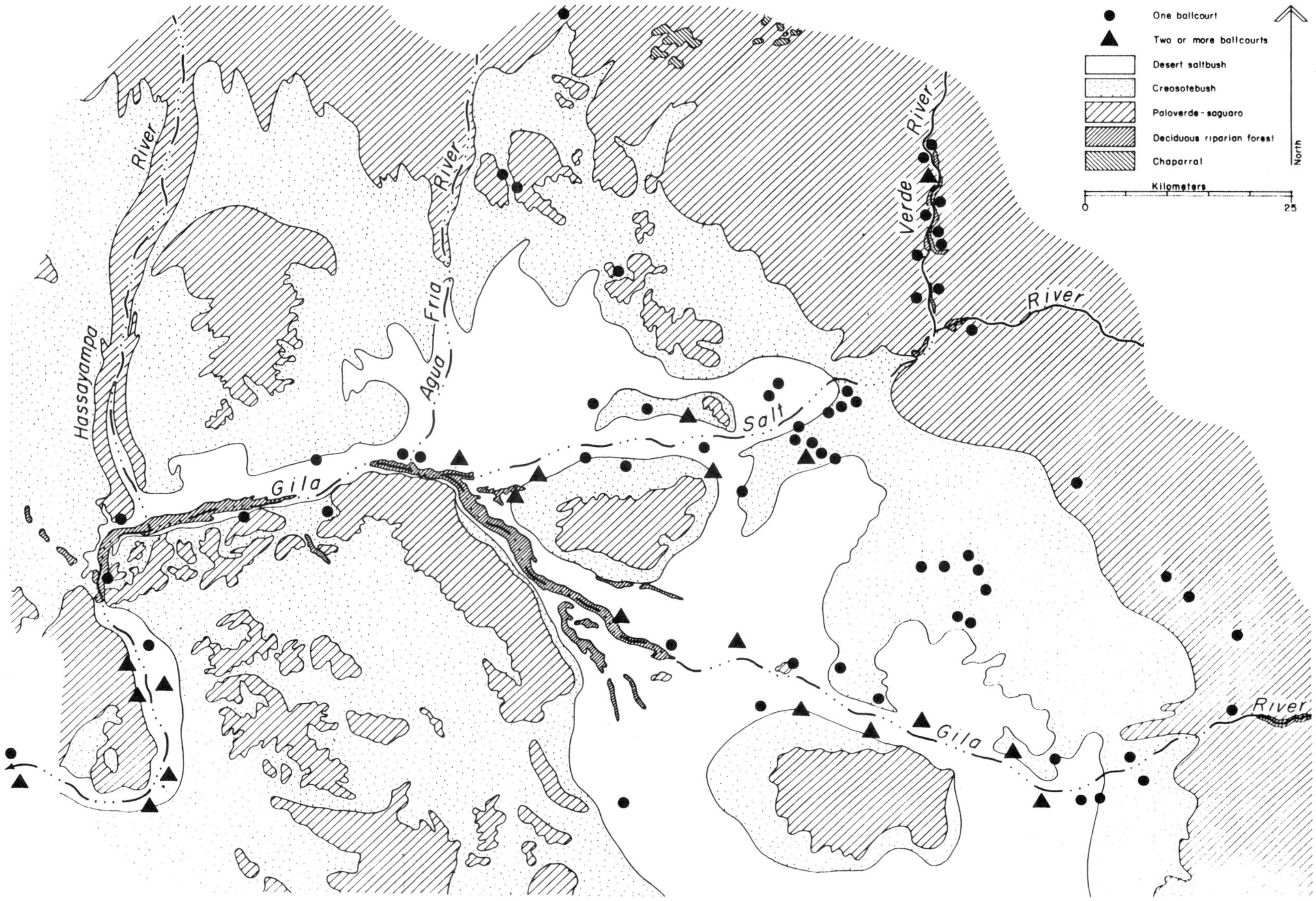

Figure 9.6. Plant communities and distribution of Hohokam ballcourt sites in the Phoenix Basin (after Wilcox and Sternberg 1983:205)

however, local systems were linked by continuities in the ballcourt distribution, forming a single regional system more than 400 km north-south by 400 km east-west (Figure 9.5). Most of the flow of intercommunity exchange was probably structured by these linear networks, and the fact that they converge in the Phoenix Basin reiterates the central role of that area in the regional system.

The coarse-grained ecological relations shown in Figure 9.6 suggest great redundancy in village subsistence on the local level (Wilcox and Sternberg 1983), although it should be noted that a more fine-grained analysis on the level of plant association does reveal patterns of variability that may have been significant (Cable et al. 1985; Wilcox 1986a). In the Phoenix Basin, most ballcourt sites are near the boundary of the creosote and sagebrush communities and they lie equivalent distances from a paloverde-saguaro community. On a regional scale, greater diversity is apparent, and the structure of village distributions is more heterogeneous. The annual round of work requirements probably varied among the local systems. As social organization evolved in each area, the local trajectories seem to have increasingly diverged, resulting in the development of unique local identities. One reflection of this is the appearance of local styles of red-on-buff pottery (Masse 1982). By the end of the Sedentary period, this process of divergence may have contributed to the collapse of the regional system and to the emergence of a new array of social alliances and exchange structures.

Village economic autonomy seems unlikely. I infer that each village required a certain level of input of matter, energy, and information per annum and supplied a certain level of output to a set of other communities. The Phoenix Basin is not a huge area. Nearly all portions of the basin can be seen from the top of Tempe Butte, and since most Hohokam villages were fixed loci of human occupation for centuries, each village potentially knew a great deal about every other one across the entire area. Furthermore, in the Phoenix Basin, two or three villages were often situated on the same canal, implying a certain degree of labor sharing and joint decision making. The relatively small village size in the basin also implies intervillage contacts. Snaketown was one of the largest Sedentary period villages, yet I estimate that it never had more than 500-600 people (Wilcox and Sternberg 1983:238). It thus seems likely that most people had affinal relatives in several other villages, whom they would visit periodically to fulfill exchange obligations.

The rituals performed in each village provided occasions when obligations among relatives, friends, and clients could be discharged. At the same time, the future structure of those obligations would be defined and redefined. I suggest that by the middle of the Sedentary period, perhaps by around AD 1000, the Hohokam ceremonial system had assumed a calendrical structure in which well-defined rituals were performed at the same time each year by the sodalities inferred above.

A calendrical structure in the Hohokam ritual cycle would have been highly advantageous in terms of organization. First, the articulation of individual village schedules would have served to choreograph the movement of people back and forth among the villages of the entire region throughout the year. This would have permitted everyone to maximize the number of obligations they could engender or fulfill. Second, conflict among local competing interests would have been minimized by the calendrical system. Decisions about when large numbers of people were to be brought together became not the whim of a few leaders but an objective necessity of the symbolic order. Cooperation could thus be more easily and frequently achieved. Third, the pacing of everyday life would have been rationalized by a calendrical structure. A rhythm would have been established in interpersonal relations that would have profoundly affected the potential for organizing work parties to carry out the annual round of subsistence tasks.

Nevertheless, the advantages to one community or to one local system were not necessarily equally realized by all. Competition for prestige and for control of a greater margin in the system of exchange flows seems likely. Innovation of new ceremonies that affected the regional choreography would have been a politically effective method in this competition. Certainly innovations were made, and not all of them appeared first in the Phoenix Basin core area. The platform mound at the Gatlin site in Gila Bend (Wasley 1960) is the most dramatic example of such a noncore-area innovation. A similar feature at Snaketown is Mound 16 (Haury 1976), which is much simpler and may have been a copy rather than the original inspiration.

Innovations like this could have had a significant effect on the regional articulation of local calendrical ceremonial systems. New ceremonies meant new schedules, and new schedules meant a reallocation of the demands on people to participate, requiring that new choices be made. The Snaketown residents may have had to innovate to counteract the success of the Gila Bend system, redefining the symbolic order to accommodate the new ways in a manner that benefited their own interests. This kind of feedback process could have produced the first calendrical system and contributed to its evolution into more complex forms. Failure to maintain a common cultural identity when innovation leads to cultural divergence, however, can result in a breakdown in the articulation among local systems, producing the collapse of the regional system. This seems to have happened in the Hohokam case (Wilcox and Sternberg 1983).

CLASSIC HOHOKAM SOCIETY AND CASA GRANDE

When the continuity of the regional village network was severed at the end of the Sedentary period, the Phoenix Basin was socially isolated or circumscribed (Carneiro 1970) in a way that had not previously occurred. The rural hinterland was also abandoned at that time. In the old riverine core zone, Snaketown was abandoned, and many other villages were radically reorganized when new central precincts were established. Platform mounds went through a rapid evolution, and a rectangular form was generally adopted in about 40 villages by the middle Classic (Gregory and Nials 1985). The structure of most of these sites is poorly known, but several had a distinctive concentric structure. This is seen most clearly at Casa Grande (Wilcox and Sternberg 1983:133-143). A central precinct of large compounds surrounds a large ballcourt and a possible central plaza. A zone with small compounds surrounds this, and then a zone with house groups that required even less labor to build makes up the site periphery. Farther afield, between the major sites, rural hamlets of perhaps four or five households and single-family farmsteads are also found.

Studies of the smaller compounds (Andresen 1985; Sires 1983) indicate that they were occupied by social groups of four or five households whose common identity was expressed by the compound wall and by burial in a separate cremation cemetery. Roomblocks in the large compounds probably contained social groups comprising similar (or even fewer) numbers of households, and no compound contained many of these roomblocks (see Haury 1945). Yet the individual houses in the large compounds were much larger and contained much more storage space than those in the small compounds (Doyel 1974). Residence on top of the platform mounds began about AD 1250 (David Gregory, personal communication 1986) and probably signals the emergence of a new kind of elite (see Wilcox 1986a, 1986b for detailed discussion).

Three levels of social ranking may be proposed for the middle Classic period (AD 1250-1350): (a) an administrative elite, who lived in better houses and probably enjoyed superior access to all resources; (b) a well-to-do class enjoying secondary status; and (c) the common folk in the village peripheries and in outlying farmsteads and hamlets. The discovery of large quantities of exotic items in a nonriverine hamlet of the Civano phase (Sires 1984) suggests both redistribution by the elite and the presence of a power structure so sanctified that the possession of exotic items was no longer necessary to maintain it. The much higher frequency of inhumation burial within compounds in the central precinct at these sites, and the rare cases of inhumation in crypts, underlines the differences in social identity between the elite and the commoners (Cushing 1890; Wilcox 1986a; Wilcox and Sternberg 1981).

By the end of the Soho phase the reorganization effort had succeeded, and the Hohokam joined with other local systems in the Salado interaction sphere, thus renewing their access to exotic valuables from Mexico. Knowledge of astronomy and the ability to anticipate critical events would have been at its height by the middle Classic period, the time when Casa Grande was built and the high point in the evolution of Hohokam culture. I submit that the arguments presented in this paper increase the probability that Casa Grande was designed and used as an astronomical observatory. Elaboration of the Sedentary period calendrical ceremonial system would have moved naturally in this direction, and the articulation with the Salado interaction sphere (Doyel and Haury 1976) may have opened access to the necessary technical knowledge--if such technical expertise was not a local development. The source of Mexican exotics in Salado sites was Casas Grandes, and the requisite astronomical knowledge apparently was possessed by the elite at this site, whose site structure is the most complex in the Southwest (Di Peso 1974; Wilcox 1986c; Wilcox and Sternberg 1983:73, 75). New research designed to explore these possibilities is needed.

THE SOUTHWEST AND THE MESOAMERICAN TRADITION

The Eastern Anasazi populations in the San Juan Basin experienced an evolution in their ceremonial systems that culminated in the Pueblo systems, whose calendrical structure and similarity to Mesoamerican forms were noted above. On grounds similar to those cited for the Hohokam, it may be possible to show that calendrically structured ceremonial systems emerged in Chaco Canyon by about AD 1000 and served to integrate the Chacoan regional system. Like the Phoenix Basin, Chaco Canyon was at the crossroads of a regional exchange system. Shifts in demographic balance that reduced its central position, emergence of distinctive local systems, and competitive innovation in ceremonial structure and scheduling may have generated the collapse of the Chaco regional system by AD 1130. Unlike the Phoenix Basin, where large-scale irrigation systems made local self-sufficiency possible and where sociopolitical reorganization was temporarily successful (Wilcox 1986a, 1986b), Chaco Canyon became a peripheral zone and then was wholly abandoned by AD 1300. Thus, there may be both similarities and contrasts in the evolutionary trajectories of the Hohokam and Chacoan regional systems that future research could fruitfully investigate.

At a more fundamental level, comparisons of Southwestern regional systems raise larger anthropological issues of general interest in New World prehistory. For example, were there systematic connections between the evolution of Southwestern cultures and the evolution of those of Mesoamerica? Were Southwestern cultures little traditions in the Mesoamerican great tradition? Are similarities and differences in the ceremonial systems and astronomical knowledge among Southwestern cultures a function of their interaction in panregional systems that conveyed esoteric knowledge as well as exotic items northward from Mesoamerica (Haury 1976; Kelley 1966)?

At an even deeper level is the more general problem of unity or diversity in New World prehistory (Putnam 1895). Suppose, for example, that Clovis hunters were the first hominids to enter the New World (Haynes 1980). They may have brought with them a single language, a single cosmology, and a single culture from which nearly all other aboriginal cultures of the New World evolved (Swadesh 1962). This real possibility poses anthropological problems of great complexity. Similarities among the Southwestern cultures and between Southwestern cultures and those in Mesoamerica may derive from an original unity of New World culture and parallel evolution thereafter. Alternatively, such similarities may derive both from this possibility and from systemic interaction among these populations on a panregional scale.

I have argued in this paper that the metastructure of cosmological belief found among the historical Pueblos and the prehistoric Hohokam is quite similar to that found in Mesoamerican cultures. Secondly, I have argued that calendrical ceremonial systems emerged among the Hohokam and possibly the Chacoan Anasazi by AD 1000. These Southwestern groups appear to have diverged from other cultures in western North America and to have maintained a closer similarity to the evolving cultures of Mesoamerica. This implies participation by Southwestern cultures in the evolution of Mesoamerican civilization. The mechanisms of this interaction remain highly controversial (Mathien and McGuire 1986), but they merit the close attention of all Southwestern archaeologists. Archaeoastronomy provides one of the domains in which these anthropological issues may fruitfully be pursued.

ACKNOWLEDGMENTS

This paper is a general synthesis of four projects conducted by the Cultural Resources Management Division, Arizona State Museum, University of Arizona, for the Western Archaeological Center, National Park Service. I would like to thank Keith Anderson, Raymond H. Thompson, R. Gwinn Vivian, the late Mark Grady, and Lynn S. Teague for the support they gave to this research. Special thanks go to Emil W. Haury and David A. Gregory for their stimulation and support and to Charles

Sternberg for his superlative drafting of the figures in all of this work. I am grateful to James Judge and Marsha Bol for inviting me to celebrate the opening of their fine exhibit on the Chaco Phenomenon at the Maxwell Museum of Anthropology; it was a highly stimulating and enjoyable experience. For any errors, ambiguity, or exercises of speculation, I alone am responsible.

REFERENCES CITED

Anderson, Keith M.
1986 Hohokam Cemeteries as Elements of Settlement Structure and Change. In *Anthropology of the Desert West: Essays in Honor of Jesse D. Jennings*, edited by Carol J. Condie and Don D. Fowler, pp. 179-201. University of Utah Anthropological Papers 110. Salt Lake City.

Andresen, John M.
1985 Pottery and Architecture at Compound F, Casa Grande Ruins National Monument, Arizona. In *Proceedings of the 1983 Hohokam Symposium*, Part II, edited by Alfred E. Dittert, Jr., and Donald Dove, pp. 595-640. Occasional Papers 2. Arizona Archaeological Society, Phoenix.

Aveni, Anthony F.
1982 *Skywatchers of Ancient Mexico*. University of Texas Press, Austin.

Cable, John S., and David E. Doyel
1985 Hohokam Land-Use Patterns Along the Terraces of the Lower Salt River Valley: The Central Phoenix Project. In *Proceedings of the 1983 Hohokam Symposium*, Part I, edited by Alfred E. Dittert, Jr., and Donald Dove, pp. 263-310. Occasional Papers 2. Arizona Archaeological Society, Phoenix.
1986 Pioneer Period Village Structure and Settlement Pattern in the Phoenix Basin. In *The Hohokam Village*, edited by David E. Doyel. Ms. on file, Pueblo Grande Museum, Phoenix.

Cable, John S., Kathleen S. Hoffman, David E. Doyel, and Fran Ritz, editors
1985 *City of Phoenix Archaeology of the Original Townsite, Block 24-East.* Soil Systems Publications in Archaeology 8. Phoenix.
1970 A Theory of the Origin of the State. *Science* 169:733-738.

Chamberlin, Neil W.
1970 *Beyond Malthus, Population and Power*. Prentice-Hall, Englewood Cliffs.

Cohodas, Marvin
1974 The Symbolism and Ritual Function of the Middle Classic Ball Game in Mesoamerica. *American Indian Quarterly* 2(2):99-130.

Cushing, Frank Hamilton
1890 Preliminary Notes on the Origin, Working Hypothesis, and Primary Researches of the Hemenway Southwestern Archaeological Expedition. In *Congres International des Americanistes, Compte-rendu de la Septieme Session*, pp. 151-194. W. H. Kuhl, Berlin.

1892a The Hemenway Southwestern Archaeological Expedition. Ms. on file, TozzerLibrary, Harvard University, Cambridge, and Hayden Library, Arizona State University, Tempe.

1892b Manual Concepts: A Study of the Influence of Hand Usage on Culture Growth. American Anthropologist o.s. 5:289-317.

Di Peso, Charles C., editor

1974 Casas Grandes: A Fallen Trading Center of the Gran Chichimeca, Vols. 1-3. Northland Press, Flagstaff.

Doyel, David E.

1974 Excavations in the Escalante Ruin Group, Southern Arizona. Arizona State Museum Archaeological Series 37. University of Arizona, Tucson.

Doyel, David E., and Emil W. Haury

1976 The 1976 Salado Conference. The Kiva 41(1).

Gladwin, Harold S., Emil W. Haury, E. B. Sayles, and Nora Gladwin

1937 Excavations at Snaketown, I: Material Culture. Medallion Papers 25. Gila Pueblo, Globe, Arizona.

Green, Jesse, editor

1979 Zuni: Selected Writings of Frank Hamilton Cushing. University of Nebraska Press, Lincoln.

Gregory, David A., and Fred L. Nials

1985 Observations Concerning the Distribution of Classic Period Hohokam Platform Mounds. In Proceedings of the 1983 Hohokam Symposium, Part I, edited by Alfred E. Dittert, Jr., and Donald Dove, pp. 373-388. Occasional Papers 2. Arizona Archaeological Society, Phoenix.

Haury, Emil W.

1945 The Excavation of Los Muertos and Neighboring Ruins in the Salt River Valley, Southern Arizona. Papers of the Peabody Museum of American Archaeology and Ethnology 24(1). Harvard University, Cambridge.

1976 The Hohokam: Desert Farmers and Craftsmen. University of Arizona Press, Tucson.

Haynes, C. Vance

1980 The Clovis Culture. Canadian Journal of Anthropology 1:115-121.

Kelley, J. Charles

1966 Mesoamerica and the Southwestern United States. In Archaeological Frontiers and External Connections, edited by Gordon F. Ekholm and Gordon R. Willey, pp. 95-110. Handbook of Middle American Indians, Vol. 4. University of Texas Press, Austin.

Levi-Strauss, Claude

1966 The Savage Mind. University of Chicago Press, Chicago.

Masse, W. Bruce

1982 Hohokam Ceramic Art. In Southwestern Ceramics, A Comparative Review, edited by Albert H. Schroeder, pp. 71-106. The Arizona Archaeologist 15.

Mathien, F. Joan, and Randall McGuire, editors
1986 Ripples in the Chichimec Sea. Southern Illinois University Press, Carbondale.

Molloy, John
1969 The Casa Grande Archaeological Zone: Pre-Columbian Astronomical Observation. Ms. on file, Western Archaeological Center, National Park Service, Tucson.

Morss, Noel
1954 Clay Figurines of the American Southwest. Papers of the Peabody Museum of American Archaeology and Ethnology 49(1). Harvard University, Cambridge.

Nelson, Richard S.
1981 The Role of Pochteca System in Hohokam Exchange. Ph.D. dissertation, Department of Anthropology, New York University.

Ortiz, Alfonso
1969 The Tewa World. University of Chicago Press, Chicago.

Putnam, Frederick W.
1895 A Problem in American Anthropology. Annual Report of the Smithsonian Institution, 1895, pp. 478-486. Washington, D. C.

Sires, Earl
1983 Archaeological Investigations at Las Fosas (AS U:15:19). In Habitation Sites on the Gila River, Part V, edited by Lynn S. Teague and Patricia L. Crown, pp. 493-728. Hohokam Archaeology Along the Salt-Gila Aqueduct, Central Arizona Project, Vol. VI. Arizona State Museum Archaeological Series 150. University of Arizona, Tucson.
1984 Excavations at El Polvoron. In Prehistoric Occupation of the Queen Creek Delta, Part II, edited by Lynn S. Teague and Patricia L. Crown, pp. 221-356. Hohokam Archaeology Along the Salt-Gila Aqueduct, Central Arizona Project, Vol. IV. Arizona State Museum Archaeological Series 150. University of Arizona, Tucson.

Swadesh, Morris
1962 Linguistic Relations Across Bering Strait. American Anthropologist 64:1262-1291.

Titiev, Mischa
1944 Old Oraibi: A Study of the Hopi Indians of the Third Mesa. Papers of the Peabody Museum of American Archaeology and Ethnology 32(1). Harvard University, Cambridge.
1960 A Fresh Approach to the Problem of Magic and Religion. Southwestern Journal of Anthropology 16:292-298.

Vivian, R. Gwinn, Dulce N. Dodgen, and Gayle H. Hartmann
1978 Wooden Ritual Artifacts from Chaco Canyon, New Mexico: The Chetro Ketl Collection. Anthropological Papers of the University of Arizona 32. Tucson.

Vogt, Evon Z.
1981 Some Aspects of the Sacred Geography of Highland Chiapas. In Mesoamerican Sites and World Views, edited by Elizabeth P. Benson, pp. 119-142. Dumbarton Oaks Research Library and Collections, Washington, D. C.

Wasley, William W.
1960 A Hohokam Platform Mound at the Gatlin Site, Gila Bend, Arizona. American Antiquity 26:244-262.

Wilcox, David R.
1985 Preliminary Report on New Data on Hohokam Ballcourts. In Proceedings of the 1983 Hohokam Symposium, Part II, edited by Alfred E. Dittert, Jr., and Donald Dove, pp. 641-654. Occasional Papers 2. Arizona Archaeological Society, Phoenix.
1986a Frank Modisle's Investigations of the Site of La Ciudad. Arizona State University Anthropological Field Study, in press. Tempe.
1986b Site Structure and Maximum Extent of the Pueblo Grande Site. Ms. on file, Pueblo Grande Museum, Phoenix.
1986c The Tepiman Connection: A Model of Southwestern-Mesoamerican Interaction. In Ripples in the Chichimec Sea, edited by Joan F. Mathien and Randall McGuire. Southern Illinois University Press, Carbondale.
1986d Interim Report. Excavations of Three Sites on the Bottomless Pits Mesa, Flagstaff, Arizona. Ms. on file, Museum of Northern Arizona, Flagstaff.

Wilcox, David R., and Lynette O. Shenk
1977 The Architecture of the Casa Grande and Its Interpretation. Arizona State Museum Archaeological Series 115. University of Arizona, Tucson.

Wilcox, David R., and Charles Sternberg
1981 Additional Studies of the Architecture of the Casa Grande and Its Interpretation. Arizona State Museum Archaeological Series 146. University of Arizona, Tucson.
1983 Hohokam Ballcourts and Their Interpretation. Arizona State Museum Archaeological Series 160. University of Arizona, Tucson.

Wilcox, David R., Thomas R. McGuire, and Charles Sternberg
1981 Snaketown Revisited. Arizona State Museum Archaeological Series 155. University of Arizona, Tucson.

Chapter 10

The Nature of the Evidence:

Archaeoastronomy in the Prehistoric Southwest

M. Jane Young

In the words of this symposium's organizer, James Judge, "the purpose of this symposium [was] to synthesize recent evidence of archaeoastronomy in the New World, with particular reference to the southwestern United States, and to discuss its role in the ceremonial life of prehistoric peoples" (see Chapter 1). Like other such symposia focusing on the nature, role, and direction of archaeoastronomical studies, this one may be said to occur at a critical time (Williamson 1978; Young 1979a, 1979b). In the last few years, theoretical developments in the study of archaeoastronomy in the American Southwest have necessitated a more interdisciplinary approach than that which has been used in the past. Scholars working independently of one another have "unearthed" various phenomena that point to a well-developed system of astronomical observation used by the prehistoric inhabitants of this area, but to date, such discoveries have been treated as more or less isolated cases. In his plea that we begin to study the implications of astronomical knowledge for socio-religious life, Reyman has described this as the "I'll show you my alignment if you show me yours" syndrome (Chapter 8). It is high time for the dialogue to begin.

The participants in this symposium represent the various disciplinary backgrounds--anthropology, archaeology, and astronomy among them--that must be brought to bear on the complex task of understanding the integration of astronomy and ceremonialism in prehistory. Implicit in this symposium has been the need to analyze and conceptualize these disciplinary interrelationships in order to provide a direction for future and on-going studies. We now have the potential to paint a more comprehensive picture of the early inhabitants of the American Southwest. The dialogue has begun, but it is not over.

The various participants have presented hypotheses and raised questions that focus on the theoretical framework and methodological problems of research in archaeoastronomy for this particular region, but they have also echoed the concerns voiced at conferences held in the past ten years or so that have spanned a broader geographical area. The results of such conferences have been both encouraging and discouraging. Certainly, they have underscored the need for smaller workshops and symposia, such as this one, which allow for more focused spatial and temporal concerns.

Of central import to all these gatherings, including this symposium, has been the question of whether or not it is possible for those of us from widely divergent backgrounds to develop a standard methodology for carrying out archaeoastronomical studies. The need for such standardization arises because there are currently too many specialists from disparate fields who cannot critically evaluate one another's work. Thus, the interdisciplinary nature of the archaeoastronomical endeavor itself has been both a boon and a curse.

A review of the research to date in the Southwest reveals inconsistency in the application of theory and method to such issues as the relationship between native astronomy and ceremonialism, and the time depth involved. In fact, some of the research is characterized by what might be called an idiosyncratic or, at least, a site-specific application of theory and method. This becomes particularly problematic when scholars and others embark on an endeavor for which they have no training. A case in point might be the number of amateur anthropologists in recent years who have been "besieging" Native Americans in an attempt to lend credence to their own pet theories. Certainly, this emphasizes the need to examine what any one of us, trained as we are in different academic traditions, considers to constitute evidence of a reasonable hypothesis.

I am somewhat struck, for example, by the number of spiral petroglyphs discovered throughout the country in the last few years that are claimed to manifest various light patterns at calendrically important times, especially at the solstices. Although in many cases the actual alignments seem to be demonstrable, we often do not have enough contextual information to fully evaluate the situation. For the purpose of a concrete example, let us consider the Prestons' film presentation. Although what we were shown is visually convincing, in order to evaluate the claims made therein a serious scholar would want to know more about the research methods that have informed the Prestons' discovery. Some of the questions that arise are as follows: How many spiral images are there in the area, and how many of these seem to be involved in astronomical alignments? What is the overall site situation; has it been disturbed by environmental or other influences in the past thousand years? Have the light and shadow patterns at these sites been documented at other times of the year besides the solstices and equinoxes? What is the total pattern of light throughout the year? What were the criteria of site selection both for their study and for the film? Was site selection based on a random sample of the entire population, or were only those sites known to include certain astronomical alignments selected? What is the possibility that the actual alignments occur by chance alone? Finally, and perhaps most important, what contextual information do we have about rock art in this area and about the cultural group who supposedly produced the petroglyphs and anticipated the interaction of petroglyphs and light?

It is not my intention here to single out only the Prestons' research, but rather to suggest some of the questions that must ultimately be answered before any such research can be useful to others in the field. Except in cases involving actual archaeological excavation, it should be possible to "replicate the experiment" in archaeoastronomy--to conduct independent observations of the same phenomena at the same sites and reach similar conclusions. We also ought to be able to extend our views beyond our own initial hypotheses, using previous data as a stepping stone for the next logical conclusion or for another type of approach.

Zeilik's analysis of structures at Hovenweep and Chaco Canyon (Chapter 3) is an excellent example of this procedure. He is able to build on the research of previous investigators, as well as on a thorough search of the ethnographic literature, to carry the argument one step further. He demonstrates, as a reasonable hypothesis, that in many cases not only are confirmatory alignments present, but anticipatory ones as well, and that the anticipatory alignments are far more important with respect to the timing of events than are confirmatory alignments. Thus, his results serve to call into question the characterization of such sites as the three-slab site at Fajada Butte as calendar markers, since they apparently cannot be used for such anticipatory observations.

In the spirit of interdisciplinary endeavor, we might ask ourselves whether we have presented our data in such a way that others can honestly evaluate the results, deciding for themselves if our data support our hypotheses. And further, we must remind ourselves that no study is truly objective--that the questions we ask often determine the answers we receive, and that we are constantly "influencing" our own data, shaping the environments and peoples we study with our preconceived notions. This subjectivity of research is discussed by Jojola, who suggests in Chapter 6 that some of Ruth Benedict's writings about Zuni culture are based largely on "her own romantic brand of humanism" and that she examined this culture through the filter of her own idealistic philosophy. She was, as pointed out, an anthropologist but also a storyteller. Aveni has aptly labeled this projection of ourselves and our interests onto other peoples and other times as "astro-ethnocentrism" (Chapter 2). If we can recognize the subjective nature of our research, we may become more willing to open our theories to comment and criticism. After all, the disproving of one hypothesis should lead to the formulation of another, more accurate one; the scientific endeavor by nature is not static.

Social scientist G. H. Mead suggests that the research scientist must be aware of the dynamic nature of his or her endeavor, so that

> science becomes really quite fluid instead of being a fixed dogmatic structure of the world. It becomes a method, a way of understanding the world, so that we can act with reference to it. And the problems that arise are those involved in our conduct with reference to the world [Mead 1936:288].

This view of science should lead us to regard our individual research in a more holistic manner. Each of us may have discovered only one piece of the puzzle, but if we do not try to fit the pieces together, we cannot even approach a comprehensive picture of the people we study, a picture that includes an understanding of their location in time and space, which is what concerns us here. Again, this fitting together of ideas and theories may permit us to observe areas in which various hypotheses intersect--as has been illustrated by some of the chapters in this volume--thus, in a sense, creating overlapping lines of evidence.

THE DATA

Although the title of this volume suggests a focus on archaeoastronomy and prehistory, it is clear that the techniques of ethnoastronomy, including an understanding of the cosmology of the historical Pueblos, are also necessary in this attempt to illuminate the past. We are concerned on the one hand with a people's conceptions of the cosmos and with the many ways in which they integrated these conceptions into their lives; on the other hand, we are also concerned with the various means they developed to record, predict, and in a sense control these important phenomena. Thus, both archaeo- and ethnoastronomy have been necessary components of this volume, linked by an emphasis on the interplay of material culture and oral tradition as a key to the role of astronomy in the prehistoric Southwest.

Material Culture

Part of our data base consists of the astronomical alignments of prehistoric structures. I emphasize the word prehistoric, for it implies the need to discover one's evidence elsewhere than in the written record--in the analysis of buildings, for example, or of site plans themselves. One might say that the astronomical information embodied in the structural components of a building is analogous to that embedded in the structural elements of a myth or folk tale. Wilcox (Chapter 9), Williamson (Chapter 7), and Zeilik (Chapter 3) have utilized this method and have expanded it, demonstrating ways in which information concerning socioreligious practices contained in the ethnographic record can provide clues to the meaning and use of such alignments by those who constructed the buildings or planned the cities.

Astronomical information is incorporated not only in buildings and other structures, but in other material items as well, such as petroglyphs. Several of the symposium participants have suggested that some Anasazi rock art sites were designed so that petroglyphs interact with light and shadow in a calendrically significant way. It is indeed striking that the Prestons (Chapter 11), Sofaer and Sinclair (Chapter 4), and Williamson (Chapter 7) all discuss rays of light falling upon spirals and other configurations. There is some evidence that the Anasazi established sun-watching sites, marked by iconographic elements of an astronomical nature, throughout the Southwest, and that the historical Pueblos did so as well. In the late 1880s, Cushing (1941:40) described a sun-watching location at Zuni that was marked with rock art representations of the sun, the sacred hand, the morning star, and the crescent moon. We do not know whether the spiral symbol itself was somehow perceived as an astronomical motif, but its occurrence in connection with these special manifestations of sunlight may indicate that this was so. Among the historical Pueblos, spiral petroglyphs have been described as representing wind, water, creatures associated with water (such as serpents and snails), and the journey of the people in search of the Center (Fewkes 1892:20; Young 1982). It is important to note that the Puebloans offer a variety of interpretations for this symbol. Although other rock art images, such as concentric circles and other geometric figures, are present at these apparently calendrical sites, the spiral seems to occur most frequently.

Oral Tradition

In addition to observing the astronomical concepts embodied in such material culture items as rock art and buildings, archaeo- and ethnoastronomy also seek evidence in various aspects of oral tradition. After all, the degree of astronomical knowledge revealed by the alignments of some of the buildings in the Southwest, such as those discussed by Williamson (Chapter 7) and Zeilik (Chapter 3), suggests that information concerning the observed regular cycles of the sun, and other astronomical phenomena as well, was transmitted orally over the span of many years. Where else, then, would one look for clues to the nature of this knowledge if not in oral tradition and its interrelationship with the information embedded in the structure itself? Further, where there is evidence of a high degree of cultural continuity, one can look to relatively recent oral tradition for clues to the ways in which past cultures interpreted astronomical events.

This line of inquiry includes exploring the cultural perception of the cosmos as revealed in the poetic language of myth. Although the metaphoric nature of myth allows for many interpretations, it can function on one level as a formulaic way of remembering the recurrent cycles of celestial events. Indeed, some scholars have been able to demonstrate that myth serves as a mnemonic device in traditional astronomy.

Jojola has suggested that story telling can be used to translate the physical aspects of the world into metaphysical ones (see Chapter 6). It would seem, then, that folktales and myths should be the starting point in the search for insight into cosmological beliefs, both of the storyteller and of the audience. Jojola warns, however, that analyses by researchers from outside the cultural context have often led to little more than other kinds of stories--the stories that ethnographers have constructed about the native stories.

He further explains that celestial personifications, such as Sun Father and Moon Mother, and the relationships between these personae tend to reinforce the listeners' perception of their responsibility in the maintenance of the natural order. In this respect, Jojola's theories are consonant with those of folklorists Toelken (Toelken and Scott 1981) and Hymes (1981), both of whom have spent many years translating and interpreting Native American folktales and myths. Both have emphasized the poetic structure of such narratives, suggesting that, in addition to presenting an underlying message about the order of the universe, the narrative itself, by its very structure (e.g., measured verse), reinforces the idea of order.

Williamson makes similar statements about the importance of oral tradition to archaeoastronomy. While Jojola suggests that traditional stories embody cosmological principles, which are reinforced in the telling, Williamson uses an Acoma myth to shed light on the astronomical symbolism incorporated in Casa Rinconada. He reminds us further that the forms of the Navajo hogan and Pawnee earth lodge are also dictated by origin myths that reveal an elaborate network of astronomically related symbolic associations. Thus, both the myth and the material culture are models of the cosmos.

I am going to mention briefly my own analysis of examples of Zuni oral tradition because it relates to Jojola's ideas about Pueblo verbal art. The following example illustrates the importance of solar motions, especially the winter solstice, as an ordering principle in Zuni verbal art. I have been retranslating some of the Zuni ritual poetry collected by Ruth Bunzel in the 1930s and have found that astronomical motifs (in this case, the movement of the sun along the horizon and the appearance of the moon near the time of the winter solstice) serve as major organizational elements, not only of the ritual poetry but of the ritual itself.

My retranslation of a portion of "Sayatasha's Night Chant," the major ritual poem of the Shalako ceremony at Zuni, is presented in Figure 10.1. The Shalako is the final ceremony in the annual cycle of ceremonies of the kachina priests, and it represents the coming of the gods to Zuni (Bunzel 1932:710-756). This ceremony also signals the approach of the winter solstice, which marks the New Year for the Zuni--the time when the sun "turns around" in its journey along the horizon and a new ceremonial cycle begins. It is significant that the Sayatasha impersonator who chants this ritual poem has chief responsibility for observations of the moon, and it is he who plants the first prayer sticks at the first full moon after the close of the winter solstice ceremony.

The first part of "Sayatasha's Night Chant" is divided into five major sections by parallel lines that are strikingly similar in construction and wording; these lines describe the movements of the sun along the horizon and the appearance of the moon near the time of the winter solstice (Young 1978). The initial lines in each stanza serve to establish the framework for the rest of the lines in that stanza. In other words, the beginning lines indicate the time of year and the location in the ceremonial cycle. I have represented here only the astronomically significant initial lines of each stanza. The drawings in Figure 10.1 are a graphic representation of the verbal message; I have indented the lines to indicate parallelism and repetition. The following are brief descriptions of the meaning of the initial lines:

(I-1) The location is sunset, facing west. The time is just prior to the winter solstice. As shown in the drawing, the left-hand sacred dwelling place indicates the position along the horizon at which the sun appears to stand still during the winter solstice. "A little way to go before he could stand to center" refers to the time of Shalako, the ceremony that takes place just before the winter solstice.

(I-2) The time is sunset of the winter solstice. "Sitting down" refers to the fact that the sun appears to set in the same place for four consecutive days at the winter solstice.

(I-3) The time is just before sunset of the winter solstice. At this point the poem goes back in time to describe the ritual activities performed in preparation for the solstice.

(I-4) The time is sunset after the winter solstice. The four days during which the sun appears to "stand still" have just passed; the sun has "turned around" and will continue its journey northward along the horizon.

(I-5) The first part of this line indicates that the moon is not yet full, and therefore it is not yet time for the planting of prayer sticks. The moon's position at this time may be near sunset in the west. As the month progresses, the moon appears to slip eastward each night until it becomes full; at that time, the moon and sun appear to be opposite each other. Thus, the second part of the line depicts the first full moon after the winter solstice, the signal for the

SAYATASHA'S NIGHT CHANT

(I-1) When the sun who is our father
  hither to his left hand sacred dwelling place
    had a little way to go before he could stand to the center

(I-2) When the sun who is our father
  to his sacred dwelling place
    having gone in, sitting down,

(I-3) When the sun who is our father
  to his sacred dwelling place
    had a little way to go in before sitting down,

(I-4) When the sun who is our father
  from his sacred dwelling place
    having come out standing,

(I-5) The moon who is our mother,
  hither to the west direction
    still appearing small;
  to the east direction
    standing in the center,
      she made days into living beings.

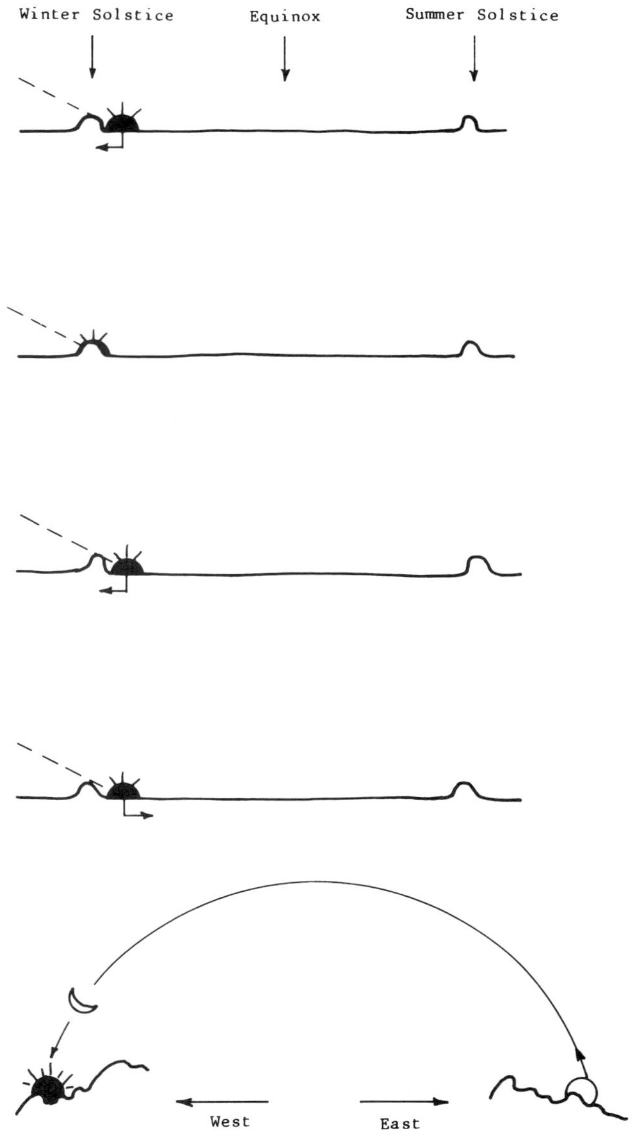

Figure 10.1. Verbal/visual depiction of astronomical elements in Sayatasha's Night Chant

> Sayatasha impersonator to plant the first prayer sticks of the year at an area with a spring.

Although I have discussed only one example, additional retranslations indicate that astronomical motifs form major organizational elements in much of the Zuni verbal art. One might say that, just as the physical structure of the kiva incorporates the motion of the sun, so does the ritual prayer uttered in the kiva.

PRECISION

It was suggested in the guidelines for this symposium that one of the specific issues to be addressed by the authors of this volume, with reference both to the astronomical practices of prehistoric peoples and to our analyses of these practices, was that of precision--an issue that has also been raised as central at other conferences focusing on archaeoastronomy. I will briefly describe some of the problems concerning precision raised in one prior instance. At at archaeoastronomy meeting in 1979, Aveni stressed the problem of twentieth-century technologists imposing their intellectual "grid" on ancient cultures. He also suggested that, in order to avoid ethnocentrism, archaeoastronomy should be more strongly connected to cultural anthropology. Others, in agreeing with Aveni's position, stressed that archaeoastronomy, especially in the Americas, involves more than merely going out into the field with equipment and determining the orientation of extant structures. They described the archaeoastronomer's role as one of trying to comprehend ancient mental systems in the context of the culture in which they developed. This led to Aveni's assertion that precision and accuracy of measurement are <u>not</u> the most important criteria in interfacing the present with the past (Young 1979b).

Aveni's earlier remarks can be used to reinforce some of the comments he has made in Chapter 2. If he is forced to repeat himself, perhaps it is because too few people are attentive to his message. Of chief importance is Aveni's suggestion that we are projecting our own concern with precision onto the people we study. As he has said, "We stand in awe when we view our ancestors in a new light--as cardboard images of ourselves in quest of the same fine detail. . . . We take the bits and pieces of ancient knowledge revealed in the cultural remains . . . [and] create an imagined system of use and acquisition of knowledge that is just like ours" (see Chapter 2). Jojola might say that we are projecting our story about precision onto a culture that has a different story.

In evaluating the nature and extent of the astronomical observations discussed in the various chapters of this volume, i.e., their degree of perceived precision and complexity, we need to know how they fit within a specific socioreligious system. This has implications for the application of the term <u>accurate</u> to any alignment or measuring device. Many of the authors have emphasized the precision of the alignments that are posited, yet precision is, of course, a culturally relative term. Further, we need to consider that concepts of time and its various manifestations vary greatly from culture to culture. Most of us, as members of Western culture, think of time as linear and measurable in discrete units, but there is no reason to believe that the early peoples of the Southwest perceived time the same way. We have imposed on time units of artificial meaning--weeks, hours, minutes, seconds, for example --in increasing degrees of precision. These are so much a part of our view of time that we have forgotten how very arbitrary they are.

When asked, "How accurate are these alignments?" one might well answer, "As accurate as they needed to be within the context of their use." I emphasize the final phrase because little other than a mechanical description is possible for an isolated site or occurrence for which the cultural context is unknown.

According to Reyman (Chapter 8), part of the power of the sun priest depended upon his ability to make reasonably accurate predictions of astronomical events, such as solstices and equinoxes. If he were to err he would be severely criticized and possibly even removed from office if adverse conditions coincided with his errors. Of course, Zeilik has pointed out in Chapter 3 that anticipatory observations would help the sun priest prevent such errors. How accurate, then, did the sun priest need to be, within the proper socioreligious context of these observations? Zeilik suggests that he would need to be accurate enough to predict the day of the solstice, within a day. One would expect that the structures from which it was possible to make anticipatory observations would reflect this degree of accuracy, but why would one expect to find anything more finely tuned? According to Parsons (1917:296-297), the smallest unit of time that had meaning for the Zunis was the day. Smaller divisions of time were introduced only with the widespread use of watches and other timepieces after the end of the nineteenth century.

Aveni's remarks seem particularly applicable to the site at Fajada Butte, as well as to some of the structures in Arizona described by the Prestons. Although the importance to the historical Pueblos and the prehistoric Anasazi of positions of the sun at solstice and equinox seems to be supported by various overlapping lines of evidence, including that of the ethnographic literature, there is little evidence to suggest that they were interested in the 18.6-year lunar cycle or that the precise determination of noon was important, contrary to Sofaer's suggestions. The station of the sun described as midday in many cases seems to have been an indicator of a meal time; this was hardly an event requiring accuracy to within a few minutes. A "noontime" ceremony did occur early in December at the Pueblo of Isleta, but it did not entail the determination of the exact moment of noon (Parsons 1932:292). A review of other examples in the ethnographic literature does not reveal that the Puebloans had a special interest in the precise moment of noon throughout the year.

As for the moon, both the ethnographic literature and contemporary practice emphasize the moon as an important deity whose phases are noted with interest. Although various records indicate that the Pueblos attempted to reconcile the yearly lunar and solar motions for calendric purposes, there is no mention of their knowledge of the 18.6-year lunar standstill cycle or of any interest in predicting eclipses. Judge has suggested that there is very strong evidence that Chaco Canyon was inhabited by a complex sociocultural group, complete with far-reaching trade networks; Wilcox has described similar complexity for the Hohokam at Casa Grande; and Reyman has demonstrated the possible existence of a more complex social framework for the historical Pueblos than has usually been posited. These descriptions should not necessarily lead us to expect complexity in all other aspects of these cultures, nor to expect that such complexity, when present, will reflect our own point of view, such as an interest in split-second accuracy.

It is no accident that the site at Fajada Butte has been called an American Stonehenge, and we recall the comment by Hawkes (1967:74) that "every age has the Stonehenge it deserves--or desires." Aveni suggests that, without some sort of support from the ethnohistoric or ethnographic literature, the attribution of marking of lunar extremes to this site, or any other in the area, seems more an attempt to project our own ideas, perhaps using research from the Old World as a model, onto this New World site (see Chapter 2). The suggestion that the boulders were intentionally set in place here and carved to create the solar phenomenon--the "sun daggers" --seems also to harken to the megalith builders of Europe. Neither the hypothesis that components of this site were used as precise markers of the lunar standstill cycle and the exact moment of noon, nor that which posits that the builders moved and carved the boulders to create these effects, seems to fit our current understanding of the prehistoric peoples of the Southwest as derived from a wide range of interdisciplinary research.

CULTURAL CONTEXT AND INTERPRETATION

Given the need to understand the cultural context of the materials with which we are working, and to interpret them in a manner that is consistent with what is known about the beliefs and interests of those who produced them, I am forced to note the small number of ethnologists who were participants in this symposium. While it is true that the emphasis here is on the prehistoric rather than the historical or contemporary occupants of this area, it is also true that several of the participants rely on data from the historical Pueblos to support their hypotheses concerning the prehistoric peoples. Although such an approach is quite reasonable in many instances, it must be practiced with care and never carried to unreasonable extremes, lest we fall into the trap of projecting the present onto the past.

My own work with the rock art in and around the Pueblo of Zuni, for example, illustrates both the possibilities of using current ethnographic data in interpreting the past and the dangers of such an approach. For a culture such as that of the Zunis, in which there has been a high degree of cultural continuity and in which the production of rock art is an on-going part of the tradition, I feel justified in seeking clues to the meanings and functions of these petroglyphs and pictographs not only in the archaeology of the area but also in the extensive ethnographic material collected over the past hundred years and in the comments of contemporary Zunis concerning their perceptions of these figures. But limitations to this approach have to be recognized and kept in mind at all times. It is quite likely, for example, that rock art is produced today for entirely different reasons than it was during prehistoric times. It is likely, too, that the meanings of some specific design elements, not to mention the elements themselves, have changed through time. For these reasons, discussions with Zunis concerning the meanings of rock art images are probably most informative with respect to images of recent date. Their interpretations of older images, although certainly important for understanding contemporary perceptions, may not reflect the original meanings of those images or the reasons for their production. In other words, there are limitations on the extent to which we can project contemporary interpretations into prehistory.

Contemporary interpretations sometimes reflect the influence of sources outside the culture, prominent among which are the very people who have attempted to study the culture. The latter have produced a sort of "distorting feedback" not only during their presence in the field but also later, through their writings. Of particular note is the relatively greater status and broader distribution that such written records may give to the interpretations of certain individuals within the culture than these might otherwise have had.

One example of this process concerns a rock art panel at a site called the Village of the Great Kivas (Figure 10.2). In the summer of 1979 a Zuni man interpreted this panel in a manner quite consistent with the comments made some 50 years earlier by Zuni workers among Frank Roberts's archaeological crew, by referring to a common Zuni folktale:

> The zigzag from the moon and star to the owl is the owl's flight during the night--when it would spy on the Navajo and then return to Zuni to report the number and location of the enemy to the Zuni war chief [Roberts 1932:151; Young 1982].

This may have been an instance of feedback, since this man had read some of the archaeological and ethnographic literature on Zuni, perhaps including Roberts's report.

A year later, however, the same man suggested that this same panel represented the "supernova explosion a long time ago" (Young 1982). In the summer of 1982 the tribal historian also gave this latter interpretation of the panel, adding that a recent visitor to the pueblo had told him about the Crab Nebula and the suggestion of some astronomers that this panel, and several other similar ones found in the Southwest, recorded the supernova explosion of AD 1054, which resulted in the formation of the Crab Nebula (Young 1982). The question I address here is whether or not a story about this site, told by non-Zunis, has influenced the interpretations given by some Zunis.

An interesting postscript to this particular example of the reinterpretive process is that some Zunis apparently associate the entire panel, rather than just the star and crescent, with this astronomical phenomenon, so that through time they could conceivably associate it with the zigzag alone. I was surprised recently to find that several Zunis identified the image portrayed in Figure 10.3 as the Crab Nebula (Young 1982).

The discoveries at Fajada Butte in Chaco Canyon and the ensuing widespread publicity in the media have led to another similar instance of reinterpretation at Zuni. Because so-called daggers of light interact with spiral petroglyphs at the Fajada Butte site, several individuals went to Zuni to inquire about the meaning and use of spiral symbols among contemporary Zunis. As a result, one of my Zuni acquaintances believes that the spirals at the Village of the Great Kivas (Figure 10.4) are bisected by streaks of light at the winter solstice--an impossibility, since the necessary configuration of nearby boulders, or any other physical feature that might produce such phenomena, is lacking. This suggestion that the panel serves as a solstice marker represents another recent change in interpretation; in 1979 this same man told me that the spirals represented the journey of the ancestors in search of the Center. This change in interpretation seems to have been sparked directly by a visitor who told this man about the site at Chaco Canyon and suggested that he might find

Figure 10.2. Pecked elements at ZRAS Site 1, Zuni Pueblo (2 by 1.8 m). Photo by M. Jane Young

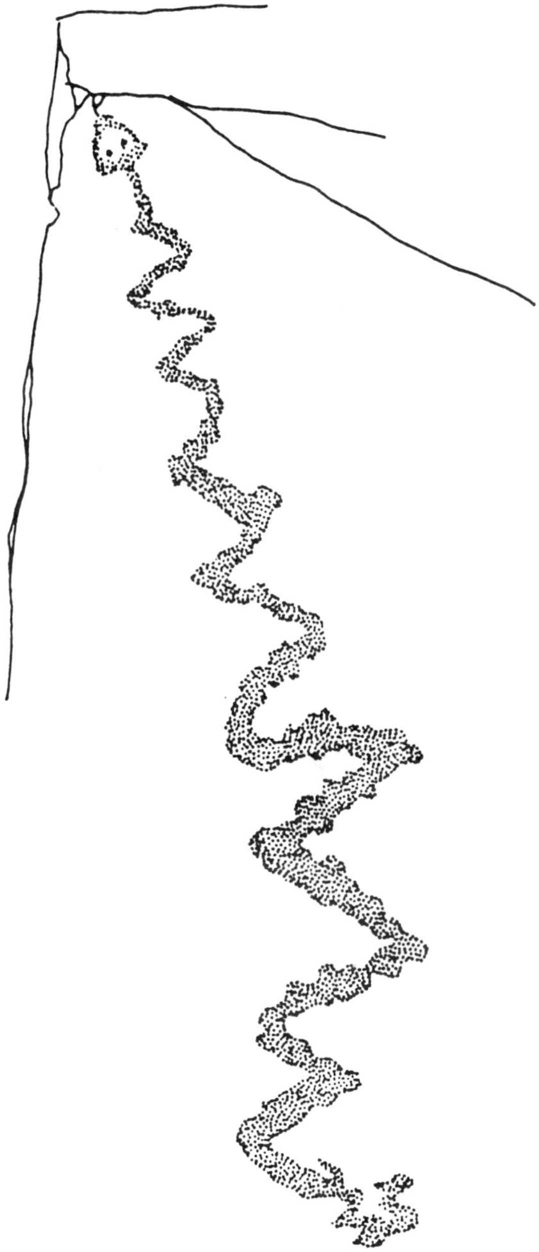

Figure 10.3. Pecked snake figure at ZRAS Site 16, Zuni Pueblo (10 by 60 cm). Drawing from slide by Murray Callahan

Figure 10.4. Pecked spirals, deer, and horned toad at ZRAS Site 1, Zuni Pueblo (1 by 0.8 m). Photo by M. Jane Young

something similar in the rock art near Zuni.

My point here is that understanding the meaning of symbols for a cultural group to which one does not belong is extremely difficult, if not impossible. One should be careful not to make sweeping generalizations and, also, not to attempt to project onto that culture a story that is really not its own. Jojola (Chapter 6) emphasizes this point as well.

I would also add that most of the Zunis to whom I talked when I was recording the rock art surrounding the pueblo were quite cautious about providing interpretations for images that they thought had been produced by their ancestors. One Zuni man described them to me as "a message from our ancestors we can no longer understand." He added, in referring to an extensive rock art site, "I don't know what it means, but I know it is important" (Young 1985:42).

These examples from Zuni have, I believe, a dual message for the archaeoastronomical endeavor. First, it would be commendable for archaeoastronomers to study the full range of anthropological literature (as well as the archaeological record) for the region in which they work. This would not only permit them to discourse properly about questions of precision but also to address questions of ritual and social function. It is vital that we familiarize ourselves with the patterns of thought and consequent actions of the people whose ancestral or contemporary material we study. This is an important means for avoiding ethnocentrism, or as Aveni said, "astro-ethnocentrism," and the validity of archaeoastronomical interpretations depends upon having a picture of the total context of the cultural area and peoples involved. That has been the purpose of this volume--to begin to put the pieces of the picture in the prehistoric Southwest together. As Judge suggested in the keynote address for the symposium (Chapter 1), archaeologists need a similar understanding of the cultural groups whose remains they study. Otherwise they too easily ascribe a ceremonial function to any perplexing object or situation they find, or conversely, they ignore the ceremonial components of the evidence entirely.

The second lesson to be drawn from the Zuni case described above concerns the dangers involved when archaeoastronomers and others with no training in anthropological fieldwork descend on the contemporary Pueblos in order to gain support for their theories about the past, an activity that could be regarded as comparable to someone without adequate training taking a shovel to an archaeological site. The result is that what is unearthed is not understood, and the context is destroyed for anyone who may come afterward. In either case, the picture is distorted.

TYPES OF ALIGNMENTS

Let us now turn to the types of alignments mentioned in various chapters of this volume, discussing them in terms of what would seem reasonable, or even intentional, in light of the ethnographic literature, and focusing particularly on social structure and ceremonialism. With the exception of Sofaer and Sinclair's suggestions of alignments to the lunar extremes (Chapter 4), most of the posited alignments are to the solstitial or equinoctial positions of the sun or to the cardinal directions. Many of these are alignments to horizon rise and set positions, but some involve other motions, such as the near-midday manifestations of light at Fajada Butte. It must also be mentioned that in addition to marking specific positions of the sun some of these sites have the potential for use as a yearly calendar, tracking the change in solar motion throughout the year.

The site at Fajada Butte has certainly been controversial, partly because we have no ethnographic context in which to interpret it. There is no real ground for proof or disproof, except perhaps for a reference to what seems to fit within the context of what we do know about the prehistoric Southwest. It is certainly consistent with known patterns of Southwestern culture to suggest that indigenous people who noticed the interesting play of light caused by the natural boulder configuration created petroglyphs at the site to serve as a sort of solar calendar. In addition, we now have evidence of other sites at which the play of light and shadow on petroglyphs corresponds with calendrically important times of the year. Williamson has mentioned one such site located near Holly House at Hovenweep National Monument (Chapter 7), and the Prestons describe 19 such sites in Arizona (Chapter 11).

That these sites might in some way mark the solstices "fits" with what we know about the calendrical and ceremonial importance of the sun for the Pueblos. As Reyman has pointed out, the sun priest had a prominent role in Pueblo society. His main function was to watch the sun's daily points of rising and setting and to announce the times for events; he was, therefore, the keeper of the calendar (Parsons 1939:122-123). This does not imply, however, that the other members of the pueblo paid no attention to the sun, nor does Reyman suggest this. Incidents have been recorded in which the sunwatcher made an error, and it was the other townspeople who ably corrected him (Forde 1931:384; Parsons 1917:296-299, 1925:74-75, 101-106). Zeilik adds another dimension to this task of sun watching with his suggestion that anticipatory observations enabled the sun priests to announce these important solar events in advance, allowing time for the necessary ceremonial preparations. Zeilik suggests, further, that it was this knowledge--"the predictive dimension of priestly power--that constituted the esoteric knowledge of the sun chief" (see Chapter 3).

Although an equinox alignment has been posited for the site near Holly House (Chapter 7), this alignment appeared to be directly related to the local planting and harvesting cycle (Williamson and Young 1979). One may also say that Williamson's recent discovery of rays of light that converge on the spiral and concentric circle petroglyphs at the summer solstice somewhat overshadows its possible role as an equinox marker (Williamson 1979). Again, this is an interpretation that is supported by the ethnographic literature, for it is the solstices that are and were of major ceremonial importance. Although the equinoxes may have been associated with planting and harvesting, there is no major ceremony that is an equinox ceremony per se.

The winter and summer solstice points were considered the most critical times of the year for the historical Pueblos, for it was then that the sun appeared to stand still for four days during its journey along the horizon. Elaborate rituals were developed and are still being performed to get the sun moving again--particularly to ensure that it "turns around" and continues its journey back along the horizon. The most elaborate of these ceremonies occurs at the winter solstice, when the sun is at its southward pause, perhaps because if the sun did not move northward again the fields would stay frozen and unsuitable for planting. The ceremony occurring at this time marks the beginning of the ritual and calendrical year for the Pueblo people (Ortiz 1972:143; Parsons 1939:570).

There is little doubt, then, concerning the central role of anticipatory solstice alignments and the more secondary role, at least with respect to the timing of events, of confirmatory solstice alignments. We might take as a future direction Aveni's suggestion that we begin to look for alignments to planetary extremes as well (see Chapter 2). Certainly the importance of the morning star is well documented in the ethnographic literature pertaining to the Pueblos. Of course, I do not suggest that this should shape the way we look at sites, but it may be that we have overlooked the possibility of such alignments in our emphasis on solar phenomena.

ASTRONOMY AND RITUAL

Let us turn briefly to Williamson's discussion of Casa Rinconada (Chapter 7) to examine the possible ritual components of alignments, i.e., ritual as it is informed by the attitudes toward time and space in a ceremonial context. Although the alignments he found at Hovenweep seem to relate to keeping the calendar, Williamson stresses that none of the alignments he has found thus far at Chaco Canyon seem to have been made for calendrical purposes. It should be noted, however, that Zeilik's examination of the windows at Pueblo Bonito indicates that they could have served to mark anticipatory alignments at the winter solstice (Chapter 3). We do not yet know enough to differentiate between the astronomical practices of the Anasazi of Hovenweep and those of Chaco Canyon, but Williamson's findings at Casa Rinconada certainly underscore the ritual component of astronomy there.

In reference to the solstice alignment from the doorway or window to the opposite wall of Casa Rinconada, Zeilik comments that in a review of the literature he found only one direct reference to calendrical sun watching from a kiva. Williamson might reply that probably the activity that took place here was not so much a calendrical sun watching as it was a ritual celebration--an inviting in of the sun at the solstice.

We must be careful not to imply too wide a gulf between the calendric and ritual components of Puebloan life, projecting our own notion of the separation of science and religion onto them. Jojola has characterized this as the problem that occurs when "attempts are made to bridge the gap between a quantitative science, such as astronomy, and the qualitative realm of anthropology." Indeed, it is virtually impossible to separate the sacred from the secular activities of Southwestern peoples, for astronomical knowledge and practices are intertwined with the way in which they perceive the cosmos. Thus, one cannot focus on the astronomy of such people without also understanding the place of astronomy in their religion or world view. The Pueblo world view is one that has been shaped by the experience of trying to support life in the harsh environment of the Southwest and by the belief that one can live harmoniously with the world of the supernatural if only one will do things in the proper ritualistic manner. Williamson suggests that the world view of the inhabitants of Chaco Canyon centered on a perception of the intimate relationship between time and space and, further, that the overall structure of Casa Rinconada was designed to unify time and space in a single symbol.

I will close with some comments based on my own fieldwork and on my study of the ethnographic literature about the perception of space and time within the context of ritual life at the Pueblo of Zuni--a perception that, I think, is characteristic of the Pueblos in general and may have been characteristic of the prehistoric peoples of the Southwest as well. What I am about to discuss is an amplification of Jojola's description of the four-dimensional model of the Pueblo world, the vertical plane of which serves to "fuse the past, present, and future."

In the Zuni ethos, time as well as space can be described as inwardly rather than outwardly directed. In Zuni mythology, as in the more general Pueblo mythology, which the Puebloans themselves regard as history, the people at the time of the beginning lived in the fourth underworld, below the surface of the earth. There was no beginning to this "time of the beginning"--it simply _was_, before the time of the emergence. Similarly, existence after death is not situated in the sky but rather inside Mother Earth, back in the time of the beginning. In the Zuni origin myth, the search for the Center after the emergence from the underworld involves geographical travels that steadily spiral inward until the Center is reached. The search for knowledge is, in a sense, synonymous with the search for the Center. The Center represents the physical existence of Zuni, the middle of any place, the time of the winter solstice, a person's heart, and that person's essence as the Center when observing the six directions or offering sacred prayer meal (Young 1985:16-18). The vehicle for attaining this knowledge, for finding the Center, is ceremonial activity.

Also apparent in Zuni ritual life is an emphasis on process rather than product--on the unity of all beings in the act of harmonious existence. For the Pueblos in general, both the moon and the sun are regarded as living beings, which are sometimes drawn into the kiva during specific rituals (Parsons 1939:378), an incorporation of the sacred into the structure itself (Williamson 1982:205, 218). Thus, space is not viewed as being outside or external to ceremonial life, nor is it regarded as something to be traveled through or overcome. Rather, such boundaries do not apply; the cosmos is one entity and the beings within it operate according to the principles of circularity rather than linearity.

This circularity applies to time as well as space. Although the Zunis may introduce a myth as having occurred "a long time ago" or "in the beginning," the events described therein are not regarded by them as being over and done with, situated at a single point in a linear flow of time; instead, they are ever-present, informing the here and now.

The Zuni view of time and space is perhaps closer to that of relativist physics--in which space and time exist as a single continuum, which is relative to the observer--than it is to the "linear perspective of the average member of Western society" (Edelman 1973:305). For the Zuni, time is cyclical; apparent in the orderly and regular motions and "returns" of the sun, moon, and stars; and organic, a continuous entity. One may even say that time for them is reversible: past, present, and future are co-existent. The Zunis believe that when people die, they return to the time and place of the myth. Many of them go to Kachina Village, which has not only mythic existence in the past but physical existence in the present at a specific lake in Arizona. Further, in the context of ceremonial activity, the gods return again and again. According to some Zunis, the kachina dancers in the plaza do not _represent_ the gods, they become the gods, and the time of the myth is one with the present. The efficacy of the rituals is due to the merging of sacred space with sacred time. Thus, one might describe the Zuni view of the universe and its dimensional aspects as a unified and orderly phenomenon. It is perhaps a similar world view that led to the construction of such structures as Casa Rinconada, dictated by the origin myth, unifying time and space, and it is in such structures that the gods reappear for the ceremonies. It may well be that the power of myth, as well as the power of astronomically oriented rituals, lies in this world view as well.

ACKNOWLEDGMENTS

My thanks to Deborah Fant, Robert H. Leibman, Ray A. Williamson, and Michael Zeilik, who read and commented on early drafts of this paper. I also thank the people of Zuni, who patiently listened to my questions about the interpretation of rock art figures surrounding the pueblo.

REFERENCES CITED

Bunzel, Ruth L.
1932 Introduction to Zuñi Ceremonialism, Zuñi Origin Myths, Zuñi Ritual Poetry, Zuñi Katchinas: An Analytical Study. In Forty-Seventh Annual Report of the Bureau of American Ethnology for the Years 1929-1930, pp. 467-1086. Smithsonian Institution, Washington, D. C.

Cushing, Frank H.
1941 My Adventures in Zuñi (reprint of 1882 edition). Peripatetic Press, Santa Fe.

Edelman, Prewitt
1973 Working Back: The New Physics and Pueblo Mythology. Southwest Review 58:302-306.

Fewkes, J. Walter
1892 A Few Tusayan Pictographs. American Anthropologist o.s. 5:9-26.

Forde, C. Daryll
1931 Hopi Agriculture and Land Ownership. Journal of the Royal Anthropological Institute 61:357-405.

Hawkes, Jacquetta
1967 God in the Machine. Antiquity 41:174-180.

Hymes, Dell H.
1981 "In Vain I Tried To Tell You": Essays in Native American Ethnopoetics. University of Pennsylvania Press, Philadelphia.

Mead, George H.
1936 Movements of Thought in the Nineteenth Century, edited by Merritt H. Moore. University of Chicago Press, Chicago.

Ortiz, Alfonso
1972 Ritual Drama and the Pueblo World View. In New Perspectives on the Pueblos, edited by Alfonso Ortiz, pp. 135-162. University of New Mexico Press, Albuquerque.

Parsons, Elsie Clews
1917 Notes on Zuñi, Part II. Memoirs of the American Anthropological Association 4:151-237.
1932 Isleta, New Mexico. In Forty-Seventh Annual Report of the Bureau of American Ethnology for the Years 1929-1930, pp. 193-466. Smithsonian Institution, Washington, D. C.
1939 Pueblo Indian Religion, 2 vols. University of Chicago Press, Chicago.

Parsons, Elsie Clews, editor
1925 A Pueblo Indian Journal, 1920-1921. Memoirs of the American Anthropological Association 32. Menasha, Wisconsin.

Roberts, Frank H. H., Jr.
1932 The Village of the Great Kivas on the Zuñi Reservation, New Mexico. Bureau of American Ethnology Bulletin III. U.S. Government Printing Office, Washington, D. C.

Toelken, J. Barre, and Tacheeni Scott
1981 Poetic Retranslation and the "Pretty Languages" of Yellowman. In Traditional American Indian Literatures: Texts and Interpretations, edited by K. Kroeber, pp. 65-116. University of Nebraska Press, Lincoln.

Williamson, Ray A.
1978 Sir Norman Lockyer Archaeoastronomy Symposium. Archaeoastronomy 1(3):4-12.
1979 Hovenweep National Monument -- Field Report. Archaeoastronomy 2(3):11-12.
1982 Casa Rinconada, Twelfth Century Anasazi Kiva. In Archaeoastronomy in the New World, edited by Anthony F. Aveni, pp. 205-219. Cambridge University Press, Cambridge, England.

Williamson, Ray A., and M. Jane Young
1979 An Equinox Sun Petroglyph Panel at Hovenweep National Monument. American Indian Rock Art, Vol. V. American Rock Art Research Association, El Toro, California.

Young, M. Jane
1978 _Translation and Analysis of Zuni Ritual Poetry_. Unpublished M.A. thesis, Department of Folklore and Folklife, University of Pennsylvania, Philadelphia.
1979a In Search of Archaeoastronomy: What Is It and Where Is It Going? _Archaeoastronomy_ 2(2):14-15.
1979b Seminar on the Role of Archaeoastronomy. _Archaeoastronomy_ 2(3):5-6.
1982 _Images of Power, Images of Beauty: Contemporary Zuni Perceptions of Rock Art_. Unpublished Ph.D. dissertation, Department of Folklore and Folklife, University of Pennsylvania, Philadelphia.
1985 Images of Power and the Power of Images: The Significance of Rock Art for Contemporary Zunis. _Journal of American Folklore_ 98:3-48.

Chapter 11

Evidence for Calendric Function at 19 Prehistoric Petroglyph Sites in Arizona

Robert A. Preston and Ann L. Preston

Thousands of prehistoric Indian rock art sites are scattered throughout the Southwest on cliffs, boulders, and cave walls (Schaafsma 1980). These images are composed of both petroglyphs (carvings) and pictographs (rock paintings). For the vast majority of sites, the meanings and functions of these drawings have been lost. Partly because of this difficulty in interpretation, the rock art sites have largely been neglected as an archaeological resource (for exceptions, see Schaafsma 1980).

During the last several years, researchers in the Southwest have observed that some of these rock art sites, all reported to be of prehistoric Anasazi origin, could potentially be used to make astronomical observations of the sun's yearly motion (Benson 1980; Ellis 1975; Sofaer et al. 1979; Williamson 1979; Williamson and Young 1979; Williamson et al. 1975, 1977). In one of these cases, a site atop Fajada Butte in Chaco Canyon, New Mexico (Sofaer et al. 1979), the interaction with the sun is particularly convincing and precise and is thus unlikely to occur as a coincidental alignment (see Chapter 4). This site consists of three large rock slabs leaning against a cliff, which allow narrow spikes of sunlight to be cast on two spiral petroglyphs on the cliff. On summer solstice, a sunlit spike moves down the cliff and precisely pierces the center of the largest spiral. The investigation of this site also indicates that winter solstice and equinox may be marked by similar sunlit images and, perhaps more speculatively, that shadows falling on the large spiral at moonrise may mark the limits of the 18.6-year lunar cycle (Sofaer et al. 1982).

The evidence for possible astronomical features in some prehistoric Southwest Indian rock art sites is indirectly supported by hints of similar rock art use at a few sites in California (Hudson et al. 1979; Rafter 1982; Romani 1981) and Baja California (Hedges 1984) and by possible purposeful astronomical alignment of prehistoric structures in both the Southwest and California (Autrey and Autrey 1981; Eidenbach 1979; Evans and Hillman 1981; Hedges 1981; Hedges 1984; Hudson et al. 1979; Molloy et al. 1973; Reyman 1971, 1976, 1977; Williamson 1977; Williamson et al. 1975, 1977; Wimberly and Eidenbach 1977). In addition, current ethnology and historical records indicate that the modern descendants of these prehistoric cultures are known to have monitored (or do monitor) the motions of the sun, moon, and stars to regulate agricultural and ceremonial calendars (Ellis 1975; Hudson et al. 1979; McCluskey 1977; Young and Williamson 1981). There is little information available on historical Pueblo usage of petroglyph sites, and no calendric functions have been reported. The lack of ethnographic support does not preclude the possible prehistoric calendric use of these sites since changes in ceremonial practices are known to have occurred in Pueblo prehistory. In addition, the spiral faded from the rock art vocabulary by approximately AD 1350 (Schaafsma 1980).

In this paper we pose the hypothesis that the apparent solar observation function of the petroglyph site on top of Fajada Butte is not unique in the Southwest but is an example of a widespread phenomenon among the Southwestern petroglyph sites that date from the same epoch. To test that hypothesis we have investigated 14 petroglyph sites within a 10 by 25 km region of the northeastern Arizona desert, plus five additional sites located at distances of 10, 25, 40, 70, and 335 km from the base sample. Eighteen of the sites appear to be of Anasazi origin, and one site is probably associated with the prehistoric Hohokam culture. We find that all of these sites can function as solar observatories, most of them in a convincing manner. Hence, it appears likely that a large number of the existing petroglyph sites in the Southwest dating from the period between AD 700 and 1300 were designed to monitor the yearly motion of the sun. Such sites have already been positively identified over a range of more than 500 km and extending over known cultural boundaries, perhaps into California.

In our study we find evidence for a specific prehistoric calendar marking four solar declinations. There is a high geographic density of solar observation sites and a high density of solar petroglyph markers per site. Among the 19 sites, 58 probable solar markers have been found, with 14 markers being identified at the most intensively studied site. Many solar markers were found to have multiple calendric purposes, with 93 sunlight interactions occurring at the 58 markers. Owing to the stability of both the local geology and the track of the sun through the sky during the roughly 1000 years since the petroglyphs were made, all of the claimed solar markers still work as they did originally, and we have watched all of them function.

METHODOLOGY

Our study has been empirical in nature. Our goal was only to ascertain if the petroglyph sites could be convincingly identified as sun-watching stations, and if so, how they functioned. We have specifically avoided speculation as to the cultural meaning and use of these sites so that the factual observational argument would not be obscured. Once the empirical case for the solar markers is well established, they can be subjected to additional archaeological and cultural examination.

SITE SAMPLE

Our base sample of petroglyph sites is located in the vicinity of Petrified Forest National Park in northeastern Arizona. Figure 11.1a shows the location of the base sample (numbered 1-14) and the locations of the five additional sites (numbered 15-19). Fajada Butte is also indicated. Figure 11.1b shows the locations of the sites in the base sample in more detail, although exact descriptions are avoided in order to protect the sites.

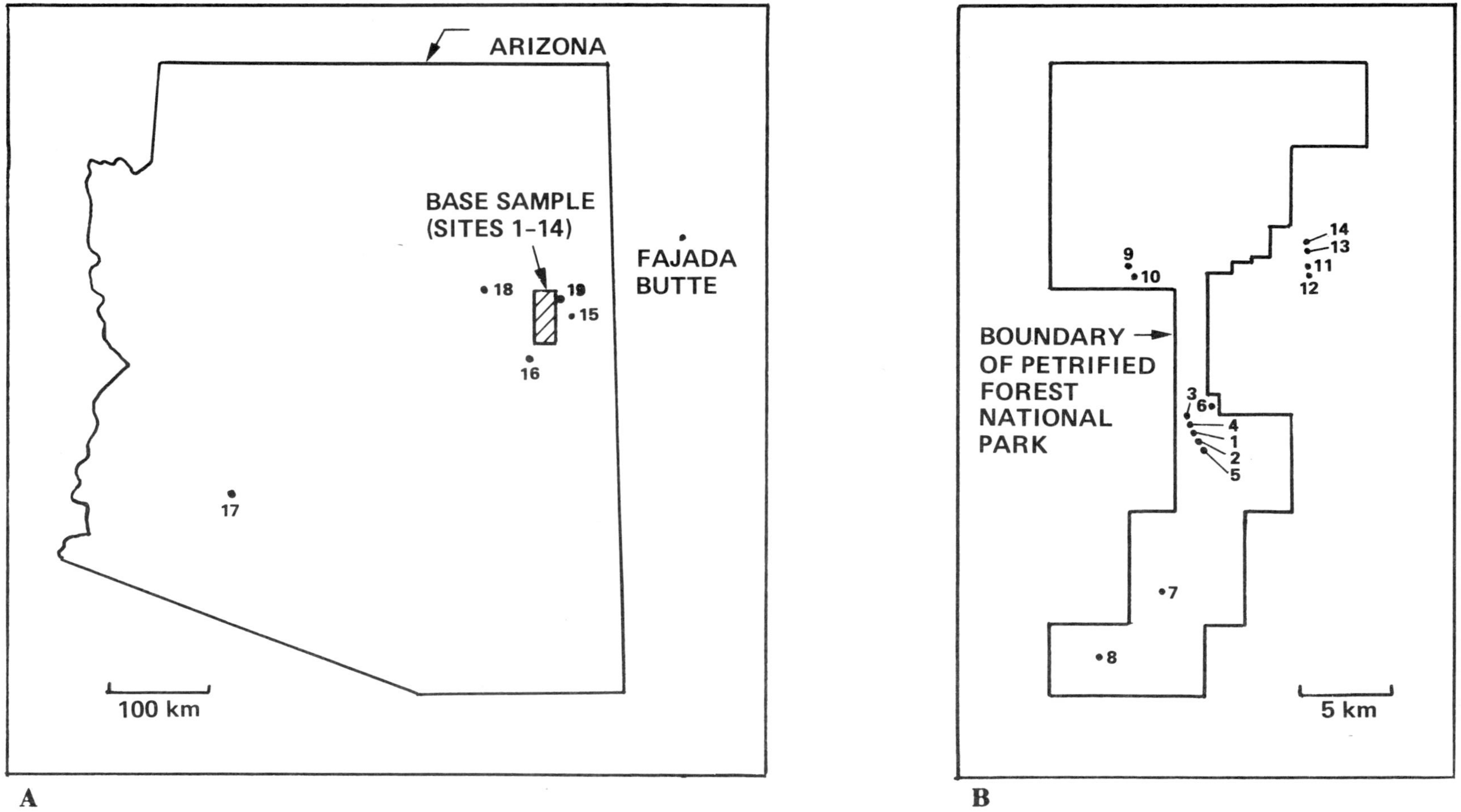

Figure 11.1. (A) Location of petroglyph sites discussed in the text
(B) Location of sites in base sample (detail of cross-hatched area in Figure 11.1a)

We defined a petroglyph site to be any group of petroglyph panels that has clearly distinct boundaries relative to any neighboring groups. This definition is important, since petroglyph sites can be closely spaced. For example, the sites numbered 1-5 are all located along a 1.5 km stretch of the same mesa edge. Site dimensions range from about 10 to 100 m, with each site containing a few to as many as tens of petroglyph panels.

We attempted to include in our base sample every major concentration of Anasazi petroglyphs belonging to the cultures known as Pueblo I, II, and III, dating roughly from AD 700 to 1300 (Schaafsma 1980; Steward 1980; Polly Schaafsma, personal communication). Most of our solar markers are probably Pueblo II and III (later than AD 900). Although there is no accurate way to date the petroglyphs, they can be crudely dated by means of stylistic traits (Schaafsma 1980).

The petroglyphs of our base sample are mainly found in the open on sandstone blocks that have broken off the tops of mesas or on small rock outcroppings. Often the jumble of blocks on the mesa slopes produces rockshelters or caves in which petroglyphs are placed. Sometimes the petroglyphs are located on cliffs formed by the original sandstone exposures.

The petroglyphs mainly consist of a large variety of geometric designs with several design elements appearing frequently (e.g., spirals, concentric circles, and wavy lines). Nongeometric figures are also present (e.g., lizards and birds). The petroglyphs were made by pecking away the oxidized coating (desert varnish) on the rocks to reveal the lighter, underlying sandstone (Schaafsma 1980).

The five additional petroglyph sites we have studied seem to date to approximately the same period as our base sample. All five contain high densities of petroglyphs. Sites 15 and 16 are situated in small canyons that cut through piñon and juniper forests. Site 18 lies in an arid region close to the Little Colorado River, while Site 19 is situated adjacent to our base sample. These sites are probably Anasazi. Site 17 is located in the Sonoran Desert region of southwestern Arizona and consists of an isolated mound of volcanic rocks. Unlike the other 18 sites, this site lies within the region occupied by the prehistoric Hohokam people and is almost certainly of Hohokam origin.

The 19 sites were studied in varying degrees of detail. Only four sites (1, 2, 3, and 17) have been intensively investigated. Although 14 solar markers displaying 30 calendric interactions have been identified at the best-studied site (Site 1), most of the other sites seem to be less complex. Six of the sites contain only a single identified marker.

TYPES OF SOLAR MARKERS

We have found evidence that the prehistoric Indians determined the occurrence of important calendric days by monitoring the sun's position with two techniques: imaging--watching sunlit images and shadows interact with petroglyphs--and sighting--direct viewing of the sun from a particular petroglyph (Hudson et al. 1979). Both methods have the same underlying principle. In order to define a line to a particular solar position one needs (a) a point of reference on the petroglyph panel, and (b) a direction to the sun from that point determined by means of one or more rock surfaces and perhaps the horizon. Sometimes it is not clear which technique was intended for a particular petroglyph, since both techniques work. The solar calendric interactions that have been found occur throughout the day, but they tend to be clustered within a few hours of sunrise or sunset, probably because there is more opportunity for rocks and horizons to intersect the sun's rays at these times.

Specific types of petroglyphs seem to have been commonly used as solar markers. Spirals and circles predominate, often with wavy lines attached. Of the 58 identified solar markers, 24 (41 percent) are single spirals, 6 (10 percent) are spirals with multiple centers, 16 (28 percent) are circles (often concentric), 9 (16 percent) are representational figures (often lizardlike), 2 (3 percent) are crosses, and 1 (2 percent) is a concentric square. The spirals, circles, and representational figures appear in many different forms. This summary of petroglyphs used for solar markers may not accurately represent the intrinsic distribution, since we have preferentially searched for spirals and circles owing to our early success with these designs. We note that the two petroglyphs identified as solar calendric markers at the three-slab site on Fajada Butte are both spirals (Sofaer et al. 1979). Ethnographic information indicates that concentric circles may be sun symbols among the modern Pueblos (Ellis 1975) and that spirals may have various interpretations (e.g., representing water, snakes, or pilgrimages; Schaafsma 1980). It has been speculated that the spiral petroglyph pattern entered the Southwest from Mexico, probably prior to AD 700 (Schaafsma 1980).

A few types of solar interactions with the petroglyphs dominate the mechanisms for calendric markings. For each solar interaction, the focal point of the petroglyph is almost always the center or the edge. The edge can be defined by the outside or inside edge of the outermost carved line or, more crudely, by the center of this line. The tip of the tail or the penis and the edge of the head or body seem to be the focal points on the representational figures.

IMAGING

Sixty-six of the 93 observed solar interactions (71 percent) involve images or shadows projected onto the petroglyphs. In these cases the interaction with the petroglyph focal points can occur in a number of forms (examples are shown in Figure 11.2). The most common form is a narrow line or spike of sunlight or shadow. For ease in terminology, we will refer to these as <u>pointers</u>, although a few do not have pointed ends. At the time of interaction with the petroglyph focal points the pointers are generally moving along the rock face roughly parallel to their long axes because of the sun's motion across the sky. The interaction of a pointer with the petroglyph focal point is almost always with the leading or trailing tip. In the case of many tangential interactions, the pointer tip appears to "brush" the edge of the petroglyph with the pointer roughly perpendicular to the line from the petroglyph center to the pointer tip. Sometimes the pointer tip traces the outside groove of the petroglyph without entering it. Sunlit dots and points of sunlit wedges (wide pointers) also interact with petroglyph focal points. The images or shadows are usually formed by rock surfaces less than 5 m from the petroglyph, since the finite angular diameter of the sun (0.54 degrees) can cause the image or shadow edge

A

B

Figure 11.2. Examples of solar imaging interactions

(A) Site 1. Forty-five days before and after winter solstice a wedge of sunlight dies out at sunset while pointing to the center of a large cross. Twenty-five different solar interactions occur at this site.

(B) Site 17. At winter solstice three thin, pointed sunlit images cross the panel. The pointer shown here intersects the center of both circles. The others run tangent to the top and bottom of both circles. Summer solstice, the equinoxes, and the days 45 days before and after winter solstice are also marked by sunlit daggers on this panel.

to become too blurred over longer projection distances.

A striking form of marking occurs when images or shadow lines suddenly appear or disappear while interacting with a petroglyph focal point. This can happen at sunset or sunrise or when the sun has reached a position such that a light path through the rocks opens or closes. In these cases, sunrise or sunset need not necessarily be associated with the distant horizon; rather, the local horizon with respect to the petroglyph panel may consist of nearby cliffs or rocks (that need not even be horizontal).

SIGHTING

Direct sighting of the sun is the second type of mechanism probably used for determining calendric dates. One need only stand at an approximate location to see the sun interact with a particular distant horizon feature on a given date. Such solar horizon watching is a known practice of some modern Pueblos. It seems likely that direct solar sighting was the most common form of prehistoric calendric observation, but such sites are difficult to identify with certainty owing to the lack of definitive evidence. We have found a particular type of sighting mechanism that seems convincing, however.

We hypothesized that the same types of petroglyphs and petroglyph focal points used for imaging interactions were used for sighting interactions (27 cases). In all instances for which we claim a probable sighting marker, the observer places his or her head against the petroglyph marker. For most of the proposed sighting interactions (74 percent), an eye is placed at the center of a circle or spiral, often so that sighting is done directly along the petroglyph rock face to its visual intersection with a distant or local horizon. When sighting is not done along the rock face containing the petroglyph, the sun appears to be aligned with a prominent feature of a distant or local horizon (for example, a notch or a hole). In all cases, the sighting alignment interacts definitively with the 0.5-degree solar disk, but no finer definition of the solar interaction has been commonly found. Although the sun is too bright for continual direct observation, we found that quick glances allowed us to identify the interaction of the sun with the sighting feature easily, especially if the sun was dimmed near sunrise or sunset.

Placement of an eye at the center of a spiral or circle petroglyph exercises quite a rigid constraint on the observer. For example, in those 10 cases in which sighting is done along the rock face containing the petroglyph, there are usually two opposing sky directions that provide a definitive sighting alignment (i.e., the interaction of the rock face with the distant or local horizons). The probability of the solar disk passing through these sky positions by chance on a particular calendar day is generally smaller than 0.008.

MULTIPLE EVENTS

One of the most convincing properties of the petroglyph markers is that some of them have multiple uses. Figure 11.3 shows that 22 percent (13 of 58) of the petroglyph markers display multiple solar interactions. In one case, two petroglyph markers on the same small rock face are each involved in seven separate solar events. Sometimes a marker is used for multiple functions on a single day, and sometimes on multiple calendar days. At two sites (Sites 1 and 7) there are a

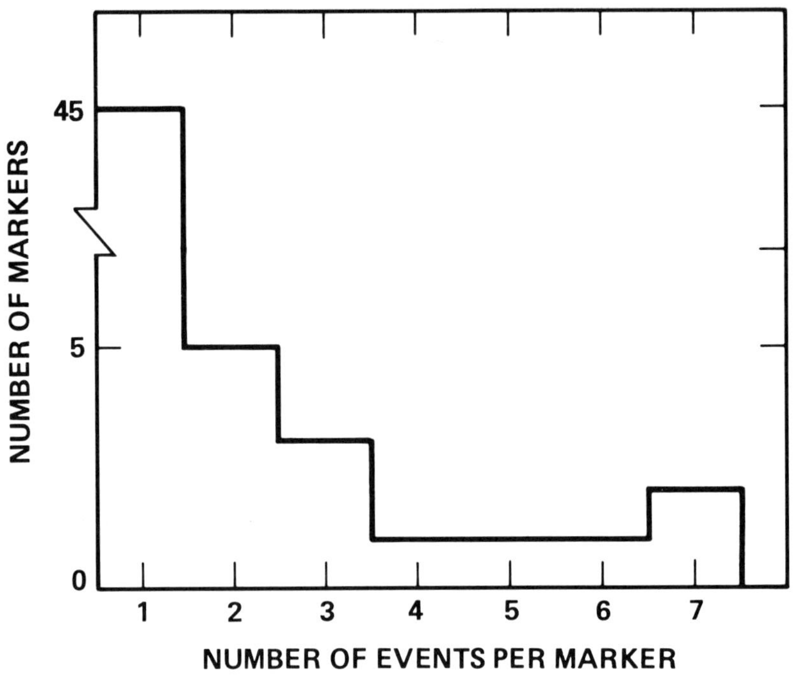

Figure 11.3. Number of calendric events associated with each of the 58 markers

combined total of 11 instances in which a single light pointer or shadow line interacts with multiple petroglyph markers on the same day. In three of these cases, the light pointer or shadow simultaneously interacts with two petroglyph markers. In one case, a single sunlit pointer interacts with five petroglyphs on the same day.

DAYS MARKED

We have found repeated and strong evidence for calendric markers associated with summer solstice and winter solstice. Two other solar declinations also appear to be marked: equinox and the period 45-48 days before and after winter solstice. Of the 93 solar interactions with petroglyphs, 39 percent were associated with summer solstice, 35 percent with winter solstice, 15 percent with the equinox, and 11 percent with the period about 45 days before and after winter solstice. We note no obvious correlation between petroglyph type and calendar day. At three sites (1, 3, and 17) all four solar declinations have been marked on the same rock face.

Subsequent to the observation of events occurring approximately 45 days from winter solstice, we noticed a possible link to the ethnology of modern tribes. From records of Hopi festivals from 1892 to 1921, a period during which the tribal calendar had not yet been disturbed by European influence, McCluskey (1977) has found that one of the most important celebrations of the modern Hopi calendar, Wuwuchim, was fixed by horizon observation of the sun "some 45 days" before winter solstice. In this first ceremony of the Hopi year, a new ritual fire is kindled and young men are initiated into the tribe, both events symbolizing new beginnings.

Estimating the accuracy of the marker interactions is a difficult problem involving many factors, including the observer's judgment. Often we were able to distinguish changes of a few millimeters in the positions of images and shadows and changes of about 0.3 solar diameters in the position of the sighted sun. In each case this involves a typical change of approximately 0.2 degrees, corresponding to about a week at the solstice and to about one day approximately 45 days from winter solstice.

It is possible that the positions of the moon, planets, and stars were also marked. Ethnological studies of Southwestern tribes show that the phases of the moon strongly influenced yearly calendars and that Venus (the morning and evening star) was important in their mythology. In general, we have not looked for such associations in the rock art. We did, however, monitor several sites at the time when the sun was at the same declination as one of the limits of the 18.6-year lunar cycle (August 1-2, 1982) without positive results. We note that the negative results produced by observations carried out on these randomly selected days with respect to the yearly motion of the sun support the argument that these glyphs served as solar calendrical markers.

CONSTRUCTION OF THE SOLAR OBSERVATORIES

Most of the solar markers seem to be associated with undisturbed natural rock formations. It is possible that the only human modification of the rocks at all the sites was the carving of the petroglyphs. For the solar devices with individual petroglyphs that serve multiple functions throughout the year, enormous care and great amounts of time were required in the carving of the petroglyphs. In these cases, the images and shadows that fell on a rock face on different calendar days had to be carefully monitored and probably sketched on the rock face with a temporary medium (e.g., paint) before the petroglyphs were carved.

POSSIBLE CONNECTION TO ROCK ART SITES IN CALIFORNIA

Solar calendric interactions have been reported at several pictograph sites in southern California (Hudson et al. 1979; Romani 1981) and northern Baja California (Hedges 1984). Two recently discovered sites in this region (Rafter 1982; Young et al. n.d.) along with another previously reported site demonstrate consistent and precise imaging mechanisms, which argues against chance occurrence. A short description of each of these sites is given in Table 11.1. Circular symbols are involved at all three sites, and sunlit images commonly interact with the center or edges of the circles.

The strong similarities between these sites and the sites we have found in Arizona are readily apparent. Speculation about possible ties between the prehistoric cultures of the Southwest, California, and northern Mexico should be undertaken with some caution, however, since these cultures are not only separated in distance but probably in time as well.

CONCLUSIONS

The use of a large sample of petroglyph sites is a valuable tool in demonstrating the credibility of calendric function in prehistoric petroglyphs. All 19 sites investigated displayed the calendric phenomena. The geographic density of sites and the density of solar markers at each site argue against the solar interactions being chance occurrences. There is a strong commonality among the solar markers, both among and within sites. These common characteristics include the petroglyph types, the types of solar interactions, and the days marked. To strengthen these arguments further, we are presently engaged in a study that will lead to empirical estimates of the probability of chance occurrence of the precise sunlight-petroglyph interactions we have observed.

The consistency of these findings not only provides solid clues to potential functions of some of the petroglyphs that occur so frequently throughout the Southwest, but opens up the possibility of a previously untapped wealth of archaeological data. We have only just scratched the surface with our present discoveries. A much more complex and illuminating puzzle may await solution. Hundreds or even thousands of astronomical petroglyph sites may eventually be found. These sun-watching sites present one of the few available lines of empirical evidence concerning the ceremonial life of the prehistoric Indian cultures. Continued study of the astronomical functions of the petroglyph sites provides an unusual opportunity to understand the thought processes of the early inhabitants of the Southwest and their attempts to understand the order of their universe.

Table 11.1. Pictograph sites in southern California and northern Baja California

Site	Reference	Location	Culture	Dates of Sunlight Interaction	Description
Sun Cave	Young et al. n.d.	Imperial County, California	---	Winter solstice	Dot of light appears tangent to large circle and moves away; sunlit pointer runs through center of rayed circle with trailing tip
				Winter solstice ± 45 days	Triangular image vanishes with point tangent to circle
Spring Shelter	Rafter 1982	Riverside County, California	Probably Luiseño	Summer solstice	Sunlit spots form on both tips of incomplete ring; they then lengthen and join
				Winter solstice	Triangular sunlit image moves tangent to disk; thin sliver of light then shoots from triangle and bisects disk
La Rumorosa	Hedges 1984	Northern Baja	Kumeyaay	Winter solstice	First light falls on circle; tip of sunlit pointer strikes head of figure

REFERENCES CITED

Autrey, N. E., and W. R. Autrey

1981 Zodiac Ridge. In Archaeoastronomy in the Americas, edited by R. A. Williamson, pp. 81-99. Ballena Press, Los Altos, California.

Benson, C.

1980 Anasazi Sun-Watching Stations. El Palacio 86:5-9.

Eidenbach, P. L.

1979 Wizard's Roost and Wally's Dome: Continuing Investigations of Prehistoric Observatory Sites in the Sacramento Mountains, New Mexico. In Jornada Mogollon Archaeology: Proceedings of the First Jornada Conference, edited by P. Beckett and R. Wiseman, pp. 103-105. New Mexico State University, Las Cruces.

Ellis, Florence Hawley

1975 A Thousand Years of the Pueblo Sun-Moon Calendar. In Archaeoastronomy in Pre-Columbian America, edited by Anthony F. Aveni, pp. 59-87. University of Texas Press, Austin.

Evans, J. H., and H. Hillman

1981 Documentation of Some Lunar and Solar Events at Casa Grande, Arizona. In Archaeoastronomy in the Americas, edited by R. A. Williamson, pp. 133-135. Ballena Press, Los Altos, California.

Hedges, K.

1981 Winter Solstice Observatory Sites in Kumeyaay Territory, San Diego County, California. In Archaeoastronomy in the Americas, edited by R. A. Williamson, pp. 151-156. Ballena Press, Los Altos, California.

1984 Archaeoastronomical Sites in the Territory of the Kumeyaay Indians of Southern California and Northern Baja California. In California Indian Astronomy -- Visions of the Sky, edited by T. Hudson and J. Carlson, in press. Acoma Books, Ramona, California.

Hudson, T., G. Lee, and K. Hedges

1979 Solstice Observers and Observatories in Native California. Journal of California and Great Basin Anthropology 1:39-63.

McCluskey, S. C.

1977 The Astronomy of the Hopi Indians. Journal of the History of Astronomy 8:174-195.

Molloy, J. P., R. E. White, T. P. Culhert, and D. H. Kayser

1973 The Case Grande Archaeological Zone Pre-Columbian Astronomical Observatory. Manuscript.

Rafter, J.

1982 Primitive Solar and Stellar Observations in Riverside County. La Pintura 8(4), 9(1):10-11. American Rock Art Research Association, El Toro, California.

Reyman, J.
1971 Mexican Influence on Southwestern Ceremonialism. Ph.D. dissertation, Southern Illinois University. University Microfilms, Ann Arbor.
1976 Astronomy, Architecture, and Adaptation at Pueblo Bonito. Science 193:957-962.
1977 Letter to the editor. Science 79:12.

Romani, J.
1981 Astronomy and Social Integration: An Examination of Astronomy in Hunter and Gatherer Society. M.A. thesis, Department of Anthropology, California State University at Northridge.

Schaafsma, P.
1980 Indian Rock Art of the Southwest. University of New Mexico Press, Albuquerque.

Sofaer, A., R. M. Sinclair, and L. E. Doggett
1982 Lunar Markings on Fajada Butte, Chaco Canyon, New Mexico. In Archaeoastronomy in the New World, edited by Anthony F. Aveni, pp. 169-181. Cambridge University Press, Cambridge, England.

Sofaer, A., V. Zinser, and R. M. Sinclair
1979 A Unique Solar Marking Construct. Science 206:283-291.

Steward, Y. G.
1980 An Archeological Overview of Petrified Forest National Park. Publications in Anthropology No. 10. Western Archeological Center, National Park Service, Tucson.

Williamson, R. A.
1977 Letter to the Editor. Science 79:12.
1979 Some Observations on Archaeology and Archaeoastronomy. Archaeoastronomy 2:11.

Williamson, R. A., and M. J Young
1979 An Equinox Sun Petroglyph Panel at Hovenweep National Monument. American Indian Rock Art 5:71. American Indian Rock Art Research Association, El Toro, California.

Williamson, R. A., H. J. Fisher, and D. O'Flynn
1977 Anasazi Solar Observatories. In Native American Astronomy, edited by Anthony F. Aveni, pp. 203-217. University of Texas Press, Austin.

Williamson, R. A., H. J. Fisher, A. F. Williamson, and C. Cochran
1975 The Astronomical Record in Chaco Canyon, New Mexico. In Archaeoastronomy in Pre-Columbian America, edited by Anthony F. Aveni, pp. 33-43. University of Texas Press, Austin.

Wimberly, Mark, and Peter L. Eidenbach
1977 Inventory of the Cultural Resources: Sierra Blanca Ski Area Land Transfer. Human Systems Research, Tularosa, New Mexico.

Young, L. E. , E. E. Young, and R. Preston
n.d. Manuscript in preparation.

Young, M. J., and R. A. Williamson
1981 Ethnoastronomy: The Zuni Case. In Archaeoastronomy in the Americas, edited by R. A. Williamson, pp. 183-191. Ballena Press, Los Altos, California.

Chapter 12

Science, Society, Objectivity, and the Astronomies of the Southwest

Stephen C. McCluskey

We shall here call "astronomy" only those parts of human interest which are amenable to mathematical treatment. Cosmogony, mythology, and applications to astrology must be distinguished as clearly separated problems.

Otto Neugebauer
Astronomy and History

The famous astrologer Dr. John Dee once used an Aztec obsidian mirror to see into the future. We may look down our noses at his ideas, but one may be sure that in outlook he was far closer to a Maya priest astronomer than is an astronomer of our century.

J. E. S. Thompson
"Maya Astronomy"

How . . . ought one . . . produce an interpretation of the way a people lives which is neither imprisoned within their mental horizons, an ethnography of witchcraft as written by a witch, nor systematically deaf to the distinctive tonalities of their existence, an ethnography of witchcraft as written by a geometer.

Clifford Geertz
Local Knowledge

We have increasingly come to realize the richness, complexity, and even accuracy of the understandings of nature developed by "primitive" peoples. Recent studies of early astronomies have been exceptionally illustrative in this regard. Yet there is a strange reluctance to apply the words science or scientific to these intellectual activities.

An example of this reluctance appears in a brief paper on Hopi sun watching by Mischa Titiev (1960). In this paper Titiev suggests that the regularity of an astronomically based festival calendar is the crucial factor for distinguishing between religion and magic. Titiev never considers the possibility that the observational and intellectual activity involved in determining this calendric regularity might fall under the rubric of science rather than either magic or religion. He describes calendric rites as "part of every primitive society's non-empirical method for trying to gain desired ends" (Titiev 1960:298; emphasis added). This dismissal of Hopi observational activity as non-scientific is representative of a whole range of viewpoints found not only among anthropologists but also among many Hopi, who see a great gap separating the activities of the tawa-mongwi and those of the Anglo astronomer. It is my contention that the widespread uncritical acceptance of this perceived separation seriously impoverishes our understanding of "primitive" cultures.

Several reasons have been offered for dismissing "primitive science," probably the most general being that it is static and unchanging. There is an unwillingness to consider theoretical alternatives, a failure to meet the criteria of openness and progress that are often held to be characteristic of science. A related observation, a factor that is sometimes seen as causing this lack of openness, is that "primitive science" is not separated from its cultural context. Traditional societies are perceived as linking their "ideas about astronomy to other key aspects of day-to-day life and spiritual beliefs," while modern astronomers are described as having a "self-contained spirit," an "urge to pursue new knowledge of the cosmos for its own sake" (Hadingham 1984:29).

This linkage of astronomy to society in traditional science expresses itself in several ways, the most crucial of which is the close involvement of traditional astronomies with religion. Traditional astronomies are practiced as part of ritual, and their astronomical concepts are expressed in sacred terms. Besides this involvement with religion, traditional astronomies and cosmologies often serve as models for social structures. Scientific astronomy, it is maintained, does not concern itself with religion or with society.

The astronomies of primitive peoples are also distinguished from genuine science in that they serve practical aims: the establishment of a religious or agricultural calendar being the chief one. This position assumes that real science is an abstract search for understanding, while applied science is a distinctly different, and presumably inferior, activity.

Others who have drawn this distinction between traditional and modern astronomy insist that an understanding of nature must be expressed in mathematical formulations before it can be termed science. Thus, only cultures that have developed mathematical astronomy can be described as having a science of astronomy worthy of the name (Ellegård 1981).

Anthony Aveni (1981:S12-S14) has maintained that the term science, as applied to astronomy, must inevitably mean modern western European science. Application of this term seems to divorce traditional astronomies from other elements of the culture of which they form an integral part; such "astronomies" would be artificial scholarly creations and distort a cultural unity into a form unrecognizable to those who originally employed them. Furthermore, such a separation of astronomy from the rest of the culture tempts us to evaluate these astronomies according to the norms of modern science. Because traditional astronomies may be found "inferior" relative to modern standards of observational precision or theoretical sophistication, Aveni implies that any attempt to apply modern Western standards would be a disservice to our understanding of these systems of astronomy.

Besides the reverse ethnocentrism hidden in the assumption that these astronomies would not measure up to our standards, these attitudes reflect a shared perception of the nature of science and of the scientific method that should be examined. Since classical antiquity, from Ptolemy to Kepler to the latest computer-generated ephemerides, astronomy has provided the Western world with an ideal of what a predictive science should be. Careful observations, exact predictive models, tests of the models' predictions using independent observations, and revisions of the models in the light of these observations were the chief ingredients of this view of science. Above and beyond these details of methodology was the more general premise of the disinterested scientist, who is able to "totally divorce astronomical phenomena . . . from social and religious contexts" (Aveni 1981:S13) and, for the most part, from scientific and philosophical preconceptions as well.

When discussing traditional astronomies there is a tendency to employ this model of science as an implicit touchstone to distinguish between "scientific" astronomy and "ritual" astronomy, between true gold and fool's gold (Ellegard 1981; Hadingham 1984:29). This attitude is so widely and uncritically accepted that it is almost axiomatic. In my first published discussion of Southwestern astronomy I wrote that Hopi astronomy could best be characterized as "pre-scientific" in the sense that it is directed toward practical goals, is nonmathematical, and most importantly, expresses its conceptual constructs in "theological" rather than "theoretical" terms (McCluskey 1977:190-192). A particularly cogent criticism of this naive analysis of Hopi astronomy was recently made by Zuidema (1982:450-453), with whose general argument I agree in all but a few technical details (McCluskey 1981, 1982:55).

Reliance on the model of exact sciences described above is becoming questionable in light of recent studies in the history and sociology of science. These studies suggest that practicing scientists do not really act the way that the model suggests. If these suggestions are correct, much of the model outlined above must be taken as something of a polite fiction--the kind of inspirational story that we tell apprentices or potential patrons, but not one that is to be taken all that seriously. This critique of, or perhaps assault upon, the description of science outlined above is so wide ranging and extensive that two specific aspects of it deserve separate, if brief, consideration.

The first aspect of this criticism stresses that our observations of nature are inevitably theory laden; we do not see nature "as it really is" but rather through the filter of existing theory, and we are constrained to make our observations in terms of that theory (Hanson 1958; Kuhn 1970). Subsequent historical and sociological investigations have generally confirmed these insights into the effect of theory upon observations in actual scientific practice. Scientists in general, and astronomers in particular, do not behave as prescribed by the commonly accepted model of science.

Hetherington (1983) summarizes a number of studies concerning the effects of theoretical expectations upon modern astronomical observations. In a number of cases, respected astronomers reported measurements that satisfied then-current theoretical expectations, yet these measurements were subsequently shown to be in error. Hetherington speculates that social and personal pressures to produce results contribute to such self-deception and to the subsequent concealment of such cases from the public view of science. Nonetheless, given that scientists must choose to ask specific questions of nature rather than randomly collecting data in a Baconian fashion, Hetherington believes that "the warping of judgment by knowledge, the influence on observational reports of preconceived opinion, is inevitable" (1983:729).

This finding that theoretical expectations influence observations, even within the science that has traditionally provided the paradigm for modern science, seriously challenges the general presumption of scientific objectivity. But just as Hetherington shows that theoretical assumptions and institutional pressures influence observations and their public discussion, sociologists of science have raised a second, and more significant, issue: the influence of society, both the specialized scientific community and society at large, upon scientific theories themselves. Because observations have been found to be inadequate, both philosophically and historically, to determine the details of theories, scientists have a certain freedom in choosing which theory they will accept. Given this freedom in choosing among theories, the way is open for society to affect the choice.

Steven Shapin (1982) has reviewed more than a hundred historical case studies that provide specific examples dealing with this problem. Beginning with numerous instances of the kind of "theory-laden" observations reported by Hetherington, Shapin goes on to consider several kinds of social effects upon scientific practice. These include the vested interests of specially trained scientists who seek to preserve or advance theories that require the use of their training and oppose theories that would reduce the value of their expertise; attempts by the scientific community to define scientific methods, problems, and values so as to exclude the contributions of amateurs and other outsiders; and even the influence of dominant political, religious, and social ideas upon the very forms in which scientific theories are framed and discussed.

These studies reflect an emerging consensus among historians of science holding that observations in themselves are inadequate to explain the theories that emerge from them and that social factors do influence the decisions of scientific communities. The remaining debate centers upon the extent of this social influence in specific historical instances.

A cautionary note is in order: science is not based merely on ideology or popular vote. Observations may be influenced by theoretical expectations, but they are not made up from whole cloth. If theories are underdetermined by observations, they must nonetheless be consistent with observations (or at least with most of them). The influence of culture upon science is limited by observations and experiments, but it is clear that science is not exempt from examination as a product of culture.

OTHER SOCIETIES--OTHER SCIENCES

Once we recognize that modern science is a product of our culture, we can examine the sciences of other cultures from a new perspective. We need no longer look on a close relationship between science and society as an aberration, as something that is characteristic only of "primitive" sciences and that we moderns have outgrown. Rather, we can examine all sciences as products of cultures and consider the different ways that sciences interact with cultures. We can also examine the different institutions that arise to support scientific activities and the different kinds of sciences that arise within different cultural environments. Thus the questions raised in the study of modern sciences can be directed towards the quite different sciences of "primitive" cultures. This perspective can enrich both our understanding of the astronomies and cultures of the Southwest and our understanding of the nature and development of science itself.

By considering traditional astronomies as science, and science as a product of culture and thus an object for a sociology or anthropology of knowledge, we can begin to explore the tensions over method and evidence that are apparent in the studies of early astronomies. Here, the eclectic methodological preferences of a historian seem to have something to contribute.

If methodological pluralism is appropriate in general history, it seems even more appropriate in such hybrid fields as the history of science, archaeoastronomy, or ethnoastronomy. Most of us who are interested in early astronomies admit that the iconographer, the folklorist, the ethnographer, and the astronomer have much to learn from each other. But we are often uncertain as to how to judge each other's contributions, given the differing evidence and disciplinary approach each specialist brings to the study of early astronomies. Can we apply similar standards to a piece of rock art, a ceremonial costume, a folktale, a ritual, or an observational method? Are we left with no criteria of evidence, of truth, with no way to judge what the astronomers we study were really doing? Must pluralism mean that "anything goes"?

If we examine the methodological discussions of the various specialties that contribute to the study of early astronomies, we find not an inconsistency of standards and methods but rather a discussion in which art historians, literary critics, and historians of science are raising in their disciplines much the same kind of issues that arise in archaeoastronomy. The language of these discussions has been taken from the fields of linguistics, cultural anthropology, and intellectual history (see, for example, Elkana 1981; Geertz 1983; Summers 1981).

Two factors from these discussions appear significant. First, the differences among the various specialties contributing to the study of early astronomies seem smaller than those among the various interpretive traditions within those specialties. It would appear, therefore, that we can speak meaningfully to each other, not only on facts but even on the most subtle points of interpretation. Second, when we consider these fine points of interpretation, we may find theoretical insights in existing bodies of literature that will clarify the specific problems that arise in the practice of archaeoastronomy. Students of early astronomies are not alone, and there is no need to reinvent the wheel.

One recurring element of these discussions seems especially applicable to the concerns of archaeoastronomy. This commonality is the emphasis on context, especially cultural context, as the source of meaning in "native" modes of action and thought. One particularly attractive proposal is that this meaning is not to be found by "putting oneself into someone else's skin," since that is clearly impossible, but by trying "to figure out what the devil they think they are up to" (Geertz 1983:58). This anthropological proposal for making the native's point of view meaningful to us has obvious similarities to the historian's goal of making the past meaningful to the present.

In trying to understand the Hopi, for example, we encounter the general problem of translation. Kuhn (1970:146, 198-204) and Whorf (1956: 134-159) both maintain that language, whether general or specialized, embodies a scientific system or world view. Practitioners of two scientific systems thus live in different worlds and communicate by different languages; it is only through translation that they find it possible to discuss these worlds with each other, and then only with difficulty.

And yet, in areas as diverse as biological taxonomy (Berlin et al. 1973; Hunn 1982) and concepts of time (Malotki 1983), the natives' worlds do overlap ours; translation is not impossible. Such translations are essential if historians and ethnographers are to understand the people they study. But as translators we must remain acutely sensitive to the logical nuances that stand between the natives' concepts and our own.

At the most trivial level, there is always the possibility of simple misunderstandings when we organize our observations of the world in different ways. An excellent illustration of such simple misunderstandings is the experience of ethnographers with Hopi direction concepts. Although Alexander Stephen studied the Hopi extensively from 1881 to 1894, it was not until the year before his death that he discovered that the Hopi directions were determined by the places of sunrise and sunset at the time of the solstices (Stephen 1893). The other principal ethnographer of the Hopi, the Reverend H. R. Voth (1901), whose career overlapped that of Stephen, never did solve the puzzle of the Hopi directions and regularly translated them as north, west, south, and east. We should never forget that the literal accuracy of our translations depends on such historical accidents.

Translating the Hopi word <u>Tawaki</u> (the places where the Sun rises and sets at the solstices) merely as "solstice" does not do justice to the Hopi concept. The Hopi concept includes the notion that the Sun is an animate being, and he does not merely turn at the tropics; Tawaki is the Sun's house, where he pauses for four days to interrupt his annual journey.

Similarly, we must also understand that Tawaki is not a point on a circle defined in the abstract mathematics of the spherical geometry that has served Western astronomy since Plato and Eudoxus. Instead it is a physically real place, a specific landmark among the natural irregularities of the distant horizon. Offerings are made when the Sun pauses at his house. These offerings express the wishes of the Hopi for the Sun's continued benevolence toward them and their crops during the coming season. Tawaki, unlike the Tropic of Cancer or a declination of +23 degrees 27 minutes, is a sacred place and a sacred concept.

Shifting to the secular mode of Western science, we can see that the Sun's houses also define an entire system of coordinates. That they are not our familiar geometrical coordinates is reinforced by the fact that the Hopi directions do not meet at right angles. These directions are empirical, and the angle between them is defined by the places where the Sun passes on his annual journey.

These coordinates not only provide a reference frame for the changing of the seasons and the measurement of directions, they also serve to organize disparate elements of knowledge and ritual. Colors, birds, plants, and animals are organized in a taxonomy founded on the ordering principle of the four horizontal directions, plus the above and below (Hieb 1972:84-89). Within the context of ritual, it is clear that we are dealing not with an abstract geometrical coordinate system but rather with four sacred directions, a concept that links astronomy and ritual, cosmology and art.

Astronomy, like all science, must be grounded on observations of the natural world, and so it is with the Hopi. The system of four directions provides the framework within which Hopi sunwatchers observe the Sun's annual motion, watching the Sun's rise as he moves from his winter to his summer house and his set as he moves from his summer to his winter house.

Traditionally the Hopi maintain that the Sun rests at his house for four days, yet one educated, and perhaps somewhat skeptical, Hopi remarked that this tradition is wrong. True to the ideals of modern science he called upon his own observation as the criterion of truth. But true to his Hopi culture he had seen that the Sun did not remain at his house for four days, but for sixteen (McCluskey 1978-1980). The fact that he expressed his criticism of the established tradition in terms of that tradition itself says much about the dependence of observation on theory and of theory on culture.

But if observation is theory-laden, it certainly is not arbitrary. There is a very real sense in which we can take a Hopi sunwatcher's observations as factual. Whether we consider the Sun being at Tawaki as a matter of arbitrary convention or as a statement of fact depends upon the alternatives we are considering. Saying that the Sun is at Tawaki as opposed to saying that the sun has a declination of +23 degrees 27 minutes is a matter of convention. Our choice between the two will depend on the culture in which we were raised or the problem we are trying to examine. Saying that the Sun is at Tawaki as opposed to being at _Kwi-chala_, the point on the horizon that marks the beginning of the general planting of corn, is a matter of fact (cf. Putnam 1981: 9).

By converting that choice between factual alternatives within the Hopi framework to a similar choice between alternatives within the theoretical system of modern astronomy we can evaluate Hopi observations as statements of fact. Analyses of the regularity of the dates of festivals based on observations of the rising and setting sun have shown that the Hopi sometimes attain a precision of 10 minutes of arc in their observations, clearly a respectable value for naked-eye observation (McCluskey 1977:178-181, 1982:40-43; see also Zeilik, Chapter 3).

The Hopi are also concerned with such questions of fact. An oft-quoted passage from Crow Wing (Parsons 1925:74-75) notes the concern in the village that the sunwatcher may "miss or take on some places" where the sun rises. Yet we must note that this concern is expressed in terms of missing a discrete natural marker that indicates the time for a specific planting episode; it is not a concern having to do with the numerical terms of angular precision appropriate to a geometric coordinate system. Our angular concepts of precision may help us to understand their observations in our terms, but we must not forget whose terms we are using.

This view of astronomy as (_a_) a set of observational practices and (_b_) a system of astronomical concepts provides one important approach to our study. It tells us much about the internal logic and methods of the scientific enterprise and about how these scientific ideas relate to other ideas in the culture we are studying. But this approach, by itself, has serious limitations: it tends to ignore the social interests of members of the scientific community and the interactions within that community and between it and other groups in society. It is important to examine how science works as an institution and to consider the profound impact that social factors can have on the development and transmission of ideas.

The transmission and preservation of Hopi astronomical lore takes place within the context of religious ritual. The sacred coordinate system defined by the four directions, the details of specific astronomical observations, the relationship of these observations to planting and to the growth of crops are all a prominent part of major Hopi rituals (McCluskey 1982:46-50). Transmission through ritual has important influences on the nature of astronomies like that of the Hopi. Rituals tend to be strongly traditional and conservative, and a science transmitted by means of rituals will be strongly resistant to change.

The virtue of preserving a stable core of scientific learning is commonly accepted by most students of modern science (e.g., Kuhn 1970; Lakatos 1970:132-138). For Kuhn, science only shifts from the "normal science" of preserving and developing a particular paradigm to a search for alternatives when the scientific community is faced with an intellectual crisis. In Kuhnian terms, the science of traditional nonliterate cultures seems to be an extreme case of normal science. Thus, the stability afforded by ritual can be seen as a virtue.

If we grant the modern presuppositions that learning is valuable in itself and that transmission and preservation of learning are important social activities, then we find ourselves with an enlightening perspective on the role of ritual. Our concern with native astronomies as science may lead us to emphasize the role of ritual in transmitting and preserving knowledge rather than its role in generating social cohesion.

This stable transmission of astronomical knowledge through ritual immediately raises a problem, however: how are we to have scientific change within a system whose stability is guaranteed by the stability of ritual? The principal function of astronomical observations is to regulate the agricultural and ceremonial calendars. These calendars are exquisitely linked to local climatic conditions, providing a schedule for the planting of crops that takes into account the precise length of the growing season and the germination time of the chief crops. Since all of science is embodied within a sacred tradition, its practitioners remain relatively closed to alternatives, and thus the successful integration of astronomical observations with local and variable climatic conditions represents a paradox.

One possible solution to this paradox is indicated by the differences that often arise between and within villages as to the details of their calendars. The most common reaction in the literature to these disagreements is to dismiss them as the result of simple errors; one village's sunwatcher didn't know what he was doing (Reyman 1980:42; Titiev 1972:257). Certainly that is the impression that one gains from the partisans of the dispute, but this attribution of errors to the others is also commonly found during those debates over the fundamentals of science that emerge at times of scientific crisis (Kuhn 1970). But among traditional societies all of science is fundamental, and there are few established mechanisms for scientific change.

In such a setting, factional disagreements may provide a mechanism to maintain the calendar's precise connection with the agricultural season despite the differing microclimates of specific villages and despite long-term climatic change. One could even suggest that a group whose ritually encoded scientific concepts no longer functioned effectively would lose adherents, supporters, and political power and be supplanted by a group (or groups) with better-adapted ritual traditions. This model for scientific change suggests a new element that should be considered during investigations of social change in Pueblo cultures.

An emphasis on the scientific functions of ritual places the relationship between astronomical knowledge and members of the priestly elite in a new light. It has been pointed out that members of this elite use their knowledge of astronomy as a device to enhance their own political power (Reyman 1980), but if ritual serves as a medium for the transmission of scientific knowledge, scholarship and teaching are major social functions of the priestly class.

If we were to view scholarship and teaching as the primary function of the priestly elite--and I am by no means certain that it is primary--we would have to conclude that, rather than employing astronomy as an ornament to ritual and as a tool to gain political power, the priests employ ritual and the political power derived from it as tools to ensure the transmission and preservation of scientific knowledge. None of us are so naive as to deny that modern scientists seek personal advancement and employ political means to defend their intellectual, institutional, and individual goals. Neither would we deny that the efforts of scientists to ensure the strength and stability of professional institutions contribute both to the prestige of the scientists and to the advancement of learning. We should not be surprised, therefore, to find such mixed intellectual, personal, and political motives at work among Pueblo sunwatchers.

One further aspect of this interaction of the intellectual and the political arises from the study of astronomy in its cultural context: the convergence between structures of social organization and the structures by which the world is made intelligible. To some this is merely an empirical phenomenon (Chamberlain 1982). To others it is a reflection of some deeper understanding of the very structure of human thought (Levi-Strauss 1966). Still others see this phenomenon as the result of an alliance in which priestly elites provide ideological and ritual support for political and social structures in exchange for economic support for the scholarly activities that justify those structures (Broda 1982).

As historical investigations have shown, this phenomenon is not unique to "primitive" cultures, nor does it always present itself in terms of recurring structural models. Rather, all cultures employ their own selections from a set of recurring <u>themata</u> in religion, art, politics, and science (Holton 1973). The specific correspondences between native astronomies and the societies in which they are found and the process by which these correspondences emerge suggest further areas of investigation. Such investigations may help us to unravel major theoretical issues in the historical and anthropological investigation of the relationships between ideas and society.

LOOKING AT OTHER SCIENCES

Whether we consider Native American astronomies from the perspective of the history of astronomy, of archaeoastronomy, or of ethnoastronomy; whether we use ethnographic evidence, alignments, or rock art, we must always remember that astronomy is the means by which societies seek to make their observations of the heavens intelligible. Thus, the study of other astronomies is not a kind of astronomical antiquarianism; it is rather an examination of the efforts by which a people came to understand the world around them. And in considering these efforts we must remember that all science, modern and ancient, advanced and primitive, written and oral, is in its own way a reflection of nature in the mirror of culture. The desire to make the world intelligible takes different forms in different cultures, yet the heavens remain a constant that different cultures interpret in ways reflecting elements of themselves.

For example, Pueblo astronomies emphasize the sun, the moon, and the four directions but have very little starlore. In contrast, the Pawnee and Navajo have a detailed starlore but pay little attention to the horizon calendar (Chamberlain 1982; Haile 1947). This is a suitable reflection of their ways of life. The Pueblos have a strong tie to their villages and know the local horizon intimately. The nomadic Navajo and Pawnee have little opportunity to establish a horizon calendar, so they rely on the familiar stars to mark the changing seasons.

But these differences in turn are reflected in varying social organization. The Pawnee villages were each associated with a star, the villages of the Skidi Pawnee thus forming a great constellation on the earth, mirroring a constellation in the skies. Among the Pueblos, it is common to divide the year in halves corresponding to the Sun's arrival at his house, with different groups being responsible for sun watching, the scheduling of ceremonials and planting, and certain governmental functions in each half of the year (Chamberlain 1982:122-126; Voth 1901:152).

This superficial comparison merely suggests some of the ways astronomies generated in response to substantially identical astronomical conditions reflect the conditions of everyday life and interact with the structure of society itself. More detailed comparisons of astronomies from different cultures under the same sky must be undertaken if we are to comprehend how the ordering of nature can reflect the patterns of society.

Ideas and society are never an easy pair of terms to deal with. In 1912 George Sarton founded Isis as "an international review devoted to the history of science and its cultural influences." Notice the neat directional pattern of early twentieth century optimism: science, the progressive search for truth, influences culture, not the other way around. Sarton's schema seems almost naive now, but that's what history does for you.

Because of people like Sarton, there are now many more and different examples of scientific ideas in societies. We are beginning to understand what they have to tell us--or more accurately, we are beginning to tell our stories about them. These stories of other places and times, like the stories the Hopi tell during the winter about the time long ago before the emergence, are not just stories about the past; they tell us about ourselves as well. They tell us what it is like to do science: to join with our colleagues to find out about the universe; to find out how it works and where we fit into it; and as stories they tell us about ourselves as storytellers.

Clifford Geertz (1983:92) set our specific problem at the beginning of this paper, and it is perhaps appropriate that he should state its significance at the end:

> Science, art, ideology, law, religion, technology, mathematics, even nowadays ethics and epistemology, seem genuine enough genres of cultural expression to lead us to ask (and ask, and ask) to what degree other people possess them, and to the degree that they do possess them what form do they take, and given the form that they take what light has that to shed on our own versions of them.

REFERENCES CITED

Aveni, Anthony F.
1981 Archaeoastronomy in the Maya Region: A Review of the Past Decade. _Archaeoastronomy_ 3(_JHA_ 12):S1-S16.

Berlin, Brent, Dennis E. Breedlove, and Peter H. Raven
1973 General Principles of Classification and Nomenclature in Folk Biology. _American Anthropologist_ 75:214-242.

Broda, Johanna
1982 Astronomy, _Cosmovision_, and Ideology in Pre-Hispanic Mesoamerica. In _Ethnoastronomy and Archaeoastronomy in the American Tropics_, edited by Anthony F. Aveni and Gary Urton, pp. 81-110. Annals of the New York Academy of Sciences 385. New York.

Chamberlain, Von del
1982 _When Stars Come Down to Earth: Cosmology of the Skidi Pawnee Indians of North America_. Ballena Press, Los Altos, California, and The Center for Archaeoastronomy, College Park, Maryland.

Elkana, Yehuda
1981 A Programmatic Attempt at an Anthropology of Knowledge. In _Sciences and Cultures: Anthropological and Historical Studies of the Sciences_, edited by Everett Mendelsohn and Yehuda Elkana, pp. 1-76. Sociology of the Sciences 5. D. Reidel, Boston.

Ellegard, Alvar
1981 Stone Age Science in Britain? _Current Anthropology_ 22:99-125.

Geertz, Clifford
1983 _Local Knowledge_. Basic Books, New York.

Hadingham, Evan
1984 _Early Man and the Cosmos_. Walker, New York.

Haile, Berard
1947 _Starlore Among the Navajo_. Museum of Navajo Ceremonial Art, Santa Fe.

Hanson, N. R.
1958 _Patterns of Discovery_. Cambridge University Press, Cambridge.

Hetherington, Norris S.
1983 Just How Objective is Science? _Nature_ 306:727-730.

Hieb, Louis A.
1972 _The Hopi Ritual Clown: Life as it Should Not Be._ Ph.D. dissertation, Princeton University. University Microfilms, Ann Arbor.

Holton, Gerald
1973 _Thematic Origins of Scientific Thought: Kepler to Einstein._ Harvard University Press, Cambridge.

Hunn, Eugene
1982 The Utilitarian Factor in Folk Biological Classification. _American Anthropologist_ 84:830-847.

Kuhn, Thomas S.
1970 _The Structure of Scientific Revolutions_, second edition. University of Chicago Press, Chicago.

Lakatos, Imre
1970 Falsification and the Methodology of Scientific Research Programmes. In _Criticism and the Growth of Knowledge_, edited by Imre Lakatos and Alan Musgrave, pp. 91-106. Cambridge University Press, Cambridge.

Levi-Strauss, Claude
1966 _The Savage Mind._ University of Chicago Press, Chicago.

McCluskey, Stephen C.
1977 The Astronomy of the Hopi Indians. _Journal for the History of Astronomy_ 8:175-195.
1978-1980 Field notes in author's possession. Department of History, West Virginia University, Morgantown.
1981 Comment on Ellegard (1981). _Current Anthropology_ 22:119.
1982 Historical Archaeoastronomy: The Hopi Example. In _Archaeoastronomy in the New World_, edited by Anthony F. Aveni, pp. 31-57. Cambridge University Press, Cambridge.

Malotki, Ekkehart
1983 _Hopi Time: A Linguistic Analysis of the Temporal Concepts in the Hopi Language._ Mouton, New York.

Neugebauer, Otto
1983 _Astronomy and History._ Springer Verlag, New York.

Parsons, Elsie Clews
1925 _A Pueblo Indian Journal, 1920-1921._ Memoirs of the American Anthropological Association 32. Menasha, Wisconsin.

Putnam, Hilary
1981 Convention: A Theme in Philosophy. _New Literary History_ 13:1-14.

Reyman, Jonathan
1980 The Predictive Dimensions of Priestly Power. Transactions of the Illinois State Academy of Science 72(4):40-59.

Shapin, Steven
1982 History of Science and its Sociological Reconstructions. History of Science 20:157-211.

Summers, David
1981 Conventions in the History of Art. New Literary History 13:103-125.

Stephen, Alexander M.
1893 Letter to [J. W.] Fewkes, June 29, 1893. National Anthropological Archives File 4408(4), Smithsonian Institution, Washington, D. C.

Thompson, J. E. S.
1974 Maya Astronomy. Philosophical Transactions of the Royal Society A 276:83-98.

Titiev, Mischa
1960 A Fresh Approach to the Problem of Magic and Religion. Southwestern Journal of Anthropology 16:292-298.
1972 The Hopi Indians of Old Oraibi: Change and Continuity. University of Michigan Press, Ann Arbor.

Voth, H. R.
1901 The Oraibi Powamu Ceremony. Publications of the Field Columbian Museum, Anthropological Series 3(2):60-158.

Whorf, Benjamin L.
1956 Language, Thought and Reality. MIT Press, Cambridge.

Zuidema, R. Tom
1982 Bureaucracy and Systematic Knowledge in Andean Civilization. In The Inca and Aztec States, 1400-1800, edited by G. A. Collier, R. I. Rosaldo, and J. D. Wirth, pp. 419-458. Academic Press, New York.

Chapter 13

Issues in the Archaeoastronomical Endeavor in the American Southwest

M. Jane Young

Integral to every major archaeoastronomy conference or symposium in the last ten years has been a seminar or panel discussion by a few selected scholars whose topic has been the state of the art in this field. The ensuing conversation usually centers on a definition or a redefinition of the discipline of archaeoastronomy, a delineation of its interdisciplinary connections, a plea for a more formalized methodology, and an outline of directions for future research. The panel discussion during the Maxwell Museum symposium was no exception, and many of the topics that were raised echoed debates from previous years. A new perspective was added, however, by the inclusion in both the panel and the audience of scholars who do not regard themselves as archaeoastronomers but whose disciplinary expertise is essential to an understanding of the cultural context of the individual sites that have been the focus of archaeoastronomical research in the Southwest. Major strides toward broadening this understanding were made during interchanges between panelists and members of the Puebloan groups, whose historical and contemporary astronomical practices constitute one of the principal subjects of the discipline's research. Restricting the symposium to a specific geographical area tightened the focus of the discussion, although the final consensus was that the panelists had barely scratched the surface of the central issues.

THE PANEL

The final day of the symposium was devoted to a day-long discussion with audience participation, chaired by John Carlson. Formal members of the panel were Anthony Aveni (Department of Physics and Astronomy, Colgate University), John Carlson (Center for Archaeoastronomy, University of Maryland), Fred Eggan (Department of Anthropology, University of Chicago), Ted Jojola (Native American Studies, University of New Mexico), Robert Mark (United States Geological Survey), Steven McCluskey (Department of History, University of West Virginia), Evelyn Newman (United States Geological Survey), Jonathan Reyman (Department of Sociology, Anthropology and Social Work, Illinois State University), Anna Sofaer (Solstice Project, Washington, D. C.), David Wilcox (anthropological consultant, Tucson), Ray Williamson (Office of Technology Assessment, U.S. Congress), M. Jane Young (Department of Anthropology, University of Texas), and Michael Zeilik (Department of

Physics and Astronomy, University of New Mexico). Others who contributed to the panel were Jim Judge (Chaco Center, University of New Mexico), Ann Preston, Robert Preston (Jet Propulsion Laboratory, California Institute of Technology), Curtis Schaafsma (Laboratory of Anthropology, Museum of New Mexico), Polly Schaafsma, Eugene Sekaquaptewa (Hopi Research and Development Co.), and Alex Seowtewa (Pueblo of Zuni), as well as various persons who did not identify themselves. The disciplines represented among the panelists and the audience included anthropology, archaeology, art history, astronomy, folklore, geology, history of science, and Native American studies.

The panelists addressed themselves to the following basic topics, centered on the interrelationship of astronomy and ceremony in the prehistoric Southwest.

1. The meaning and interpretation of rock art images located at sites for which an archaeoastronomical function has been posited.

2. The limitations of ethnographic analogy and the extent of cultural continuity. In the case of Chaco Canyon, this issue focused on whether or not aspects of the modern Puebloan cultures can be projected onto the early Chacoan culture.

3. The nature of the society that built the road system and the astronomically aligned structures at Chaco Canyon. How complex or stratified was this society, and what degree of technology did these activities entail?

The problem underlying all of these concerns, the bias of the investigators, was illustrated by numerous questions. What sort of evidence do we have for the astronomical practices we infer? Is there solid evidence, or are we merely finding what we are looking for? Have we considered all of the available data, or only that which supports our theories? Are our hypotheses testable? To what degree are our data representative? Can we bridge the gap between science and ceremony? In what way do our own myths and stories illustrate the fallibility of the scientific endeavor? Is compatibility possible between professional astronomy and artistic interpretation?

ARCHAEOASTRONOMY IN THE SOUTHWEST: MODELS, MATERIALS, AND METHODS

The panel discussion was initiated by a statement prepared in advance by Aveni, which addressed the major issues of the symposium as a whole. The panel was then opened to general discussion.

Aveni focused on three basic points that had arisen in the symposium's papers and comments. First, several broadly based cultural models were suggested for Chacoan culture. Aveni contrasted Judge's model, which emphasizes the importance of ritual, with Reyman's socioeconomic model. Models such as these are necessary to an understanding of Chacoan culture; without them there is no context for a discussion of Chacoan astronomical practices. Aveni added, however, that this research has not yet developed to the point at which these various models can be synthesized, stressing that astronomers, too, disagree in their interpretation of the astronomical data. According to Aveni, one question of central importance that arises is "Whatcan we learn about theories of culture from all these data?"

Aveni's second point was that more research is needed on the Southwest-Mexican connection mentioned by Judge, not only in relation to astronomical practices and cosmology but in the realms of architecture, art, and iconography as well. In his third major point, Aveni expressed his conviction that the interplay of light and shadow at astronomically significant moments on some of the petroglyphs in the Southwest, particularly on the spirals, represents deliberate creations. To support this point, he cited evidence that has come to light in the last few years and that is discussed in the papers presented by the Prestons, Sofaer and Sinclair, Williamson, and Zeilik. He added that, perhaps in response to the public's interest in sensationalism and "dazzlement by daggers of light," the arguments supporting the existence of such sites in general sometimes inflate and overemphasize the precision of the astronomical alignments and calendrical data under investigation. Aveni urged that the research on light and shadow phenomena continue, particularly research based on empirical evidence and guided by a critical perspective and the application of scientific methodology.

Aveni reminded the panelists that debates among proponents of different disciplines, especially astronomy, archaeology, and anthropology, are not new to the archaeoastronomical endeavor, and that they should not be viewed as a confrontation but as an interaction between the natural or environmental sciences and the social sciences. The recurrent issues, however, are a result of the different methodologies and disciplinary perspectives employed by the anthropologist and the astronomer. Whereas the astronomer is a proponent of cultural transcendence, holding it to be self-evident that all humans look at the heavens, develop cosmologies, and practice observational techniques for determining the seasons, the anthropologist is a cultural relativist, regarding human behavior not as determined by the environment or by any kind of universal cosmic consciousness but rather by the different ways in which societies develop and interact. It is difficult, therefore, for the anthropologist to take seriously an approach that circumvents culture and excludes cultural data as input. Aveni concluded by suggesting that we "probably have to talk a little bit less and listen a little bit more" to what the contemporary Native American peoples of the Southwest have to say.

Concurring with Aveni's final remark, Reyman suggested that one form of "listening" could be a thorough examination of the wealth of archival materials on the historical and contemporary cultures of the Southwest. Much of this unpublished archival material amplifies the printed record, but sometimes it contradicts or modifies that record.

Reyman added that ongoing destruction of sites by vandals or through various kinds of construction projects leaves not only the archaeoastronomers but also the archaeologists and the ethnologists "on the horns of a real dilemma." On the one hand, it is absolutely necessary to research archival material in depth; on the other hand, by the time one has sufficiently learned this material so that some intelligent statements can be made, the very sites one needs to study in order to check those statements have been destroyed. Reyman concluded that (a) archaeoastronomers need to immerse themselves in the archival materials, (b) they need to recognize the ongoing destruction of the archaeological record, and (c) with respect to understanding the ethnographic context, they need to know much more about Native American peoples before they attempt field study.

In response to Reyman's comments about the unpublished archival material, Sofaer remarked that this did not take into account the unspoken material, especially the traditions of secrecy that surround sacred ritual. Seconding Aveni's queries addressed to the Native Americans attending the symposium, Sofaer requested that they reply to what had been said in the two days of the symposium, paying particular attention to the question of whether or not the "written material that we're drawing from is accurate and complete."

The discussion that followed centered on issues like those articulated by Jojola, on the observation that the materials studied by archaeologists, archaeoastronomers, and ethnologists are frequently subject to misinterpretation. None of these scholars can claim a purely objective stance. They must therefore try to be aware of their own subjectivity, particularly the effects of their own cultural experiences on their attempts to explain phenomena from other cultures. It was pointed out that the approach of archaeoastronomers is further skewed because they study only the physical universe, unmindful of the spiritual universe in which Puebloan cosmology is embedded. As Eugene Sekaquaptewa reminded the panelists, their data may be factual but the cultural meaning is missing.

Alex Seowtewa from the Pueblo of Zuni addressed the issue raised earlier concerning the socioeconomic power of the sun priest, stressing that the last sun priest at Zuni left the pueblo because of financial problems. The obligations of the sun priest, he said, were serious ("from the heart") and did not entail monetary payment. Seowtewa further remarked that, although this symposium had given him insight into the scientific side of sun watching, he was doing research with the Zuni elders to better understand such practices from a Zuni perspective. He cautioned that even the early anthropologists, such as Cushing, who spent considerable time at Zuni near the turn of the century, misinterpreted cultural data. Those who wish to understand Zuni cosmology could avoid potential misinterpretation, therefore, by sharing their findings and discussing their ideas more openly with the native peoples involved.

Sekaquaptewa focused on the word rediscovery, which had been used to characterize the archaeoastronomical endeavor. He said, "We do that all the time. It's part of the Hopi way. The Hopi legends describe some of these things." He further reminded the panelists that they may be seeing only part of the picture in trying to relate astronomical practice to Hopi culture and ritual because "the basic foundation for all of this is spiritual. . . . We live this way. We don't have curious questions about sun watching, we utilize it for our own benefit. When you live with it, as part of spiritual nature, it will benefit you, you'll understand it and utilize it to your benefit." Celestial observation, Sekaquaptewa said, is used to determine the timing of events and is an integral part of ceremonialism. In conclusion, he offered this advice to the archaeoastronomers: "I'll go back to 'rediscovery.' You're going in the right direction. Come to the reservation and ask around and you will be able to put things a bit better. . . . Then on both sides we will be able to interact on a professional level."

The remarks of both Seowtewa and Sekaquaptewa reinforced Aveni's earlier comments concerning the need for input from the Pueblo peoples into interpretations of astronomical/cosmological data. Indeed, other such interactions at this symposium made it clear that cultural understanding is, necessarily, a two-way process. An issue raised later in the day is also relevant here: contemporary Pueblo societies are not homogeneous, nor can they be regarded as being congruent to cultures of the Anasazi, who created the structures that are under investigation.

In response to some of the comments made regarding his previous description of the socioeconomic role of the Pueblo sun priest, Reyman suggested that the problem was one of semantics--i.e., the formal language in which his paper was couched may have created a misunderstanding. His intention, he said, had not been to discount the spiritual underpinning of Pueblo life, nor the Zuni sun priest's responsibility for the welfare of the village. Rather, his point was that there is an integral connection between ceremonialism and mundane activities. Therefore, even the spiritual leaders are, to some extent, involved in organizing daily life within the village--mobilizing communal labor, scheduling planting activities, etc.

Once more, the discussion turned to the subjectivity inherent in archaeoastronomical research. Addressing the issue of myth vs historical accuracy, Wilcox pointed out that members of Western culture, the dominant society, often create myths about the past in order to justify the fact of their dominance of the American Indian. The role of the social scientist, he said, is to puncture such myths, such cultural misconceptions, in order to reveal a deeper truth.

Disagreeing with this terminology, Williamson argued that the term _myth_, as used by Wilcox, implies only irrationality, falsehood, and error, whereas myths and stories can also be regarded as expressions of cultural truth, the "beliefs that we all live by." Model building, Williamson said, is a kind of storytelling; the point is to be open to new stories.

A further perspective on the role of stories was supplied by Jojola, who regarded the new stories, the concept of rediscovery so often mentioned in this symposium, as a potential political tool, for Puebloans in particular and for Native Americans in general. He suggested that an awareness among the general public of the degree of scientific knowledge inherent in traditional Pueblo practices could change the popular image of the Native Americans so that the dominant society would no longer use the word _primitive_ to describe Pueblo agricultural and ceremonial activities but would begin instead to use such terms as _accuracy_ and _precision_.

Focusing on the semantic problems created by the use of these terms, and recalling Aveni's warning that one can "overplay" one's role, McCluskey reiterated that archaeoastronomers sometimes overemphasize the value of precision to the societies they study. He also noted, however, that many Native American astronomical observations exhibit such a high degree of precision that the importance of precise observation to those cultures should not be denied.

Having explored some of the problems inherent in an attempt to understand Native American astronomy in the past and in the present, the panelists next considered the related problems of interpreting rock art in the Southwest in general (and Chaco Canyon, specifically) and determining an appropriate cultural context by which to do so.

ANALYSIS, DATING, AND INTERPRETATION OF ROCK ART AT ARCHAEOASTRONOMICAL SITES

Citing its relevance to this symposium and to the task of mapping out future directions in archaeoastronomical research, Zeilik introduced the topic of the so-called sun-watching site at Wijiji in Chaco Canyon. It contains both Navajo and Anasazi rock art, although the precise determination of which is which is still a subject of much debate. Zeilik's presentation of his research indicated that the Wijiji site could have been used as a horizon-watching site, and that it also belongs in the corpus of those sites characterized by light and shadow patterns on spiral figures at the solstices.

Curt Schaafsma countered that the site at Wijiji will remain problematic until more is known about the culture to which the rock art belongs and, hence, until at least some of the solstice-related phenomena are appropriately assigned. Stylistic studies of rock art in other areas in the Southwest, said Schaafsma, could be applied to this site to help to determine the "appropriate cultural matrix of the rock art," but without any data, research on the archaeoastronomical function of the site is greatly limited.

Concurring with Schaafsma, Williamson stated that as a basis for ongoing research in Chaco Canyon it is necessary to sort out which cultures produced which rock art as well as other items of material culture. The Wijiji site may illustrate a situation in which there was Navajo reuse of an earlier Anasazi site.

Carlson pointed out that the interpretation of rock art in its cultural context at archaeoastronomical sites was one of the main themes of this symposium. He suggested that the panel discussion focus on this issue, particularly in regard to shafts of light touching various kinds of petroglyphs and pictographs.

Aveni responded by raising the following key questions: What is the relationship of the spirals to the other rock art figures at the sites under consideration? What is the distribution of spirals with respect to the distribution of other figures? How do these images fit into the literature on rock art and rock art interpretation, specifically that dealing with rock art in the American Southwest? Has a stylistic analysis of the petroglyphs in this area been conducted to determine the cultures from which they emanate? In sum, what is the general status of knowledge about Southwestern rock art? Aveni added that some research attention should be directed toward figures other than the spirals that also occur at sites singled out as being archaeoastronomically important.

Polly Schaafsma, an expert on rock art in the Southwest, replied that data are available on stylistic analysis, relative dating, and interpretation of rock art in the Southwest and that, to some extent, more is known about the context of rock art in this area than in many other areas of North America because of the available ethnographic material. For example, many of the "rock-light-petroglyph" phenomena involve rock art that can be identified as Puebloan, and ethnographic evidence from both historical and contemporary times indicates that the Pueblos have had a keen interest in observing the interaction of light and shadow phenomena. This provides an appropriate cultural context for archaeoastronomical research at such sites. Schaafsma also emphasized that the dating of rock art is frequently established along "secondary" lines, involving relative sequences, cultural material associations, superposition, etc.

A question from the audience turned the discussion to a consideration of the meaning of rock art symbols. Although Aveni had suggested earlier that the panel's attention be directed to images other than spirals that are also associated with light/shadow phenomena (for example, snakes), interest in the ubiquitous spirals tended to dominate the discussion.

Focusing on the meaning of rock art images in general, P. Schaafsma responded that while interpretation can be aided somewhat by cultural continuity, there are limitations to such an approach. For example, when the rock art images can be identified stylistically as Puebloan, then the ethnographic literature and the commentary of modern Pueblos can shed light on interpretations of some of the figures. She cautioned that modern interpretations do not necessarily reveal, or even relate to, what was meant by the original artists because meanings may change through time. From AD 1325 on, the figures, figure complexes, and relationships of figures in Pueblo rock art are similar to those found in modern Pueblo ritual and ceremony. This provides some information as to the meaning of rock art dating to the fourteenth century or later. At the same time, the early rock art in the Southwest was also related, to some extent, to Mexican iconography, so one can look to Mesoamerica for a general sense of what these symbols meant.

Echoing his earlier comments, Aveni suggested that it would be useful to be able to discuss the images with which spirals are most frequently associated. P. Schaafsma agreed that more research should be conducted in Southwestern rock art studies on the associational patterns of figures. For example, flute players and snakes are often related, as are mountain sheep and insects; modern Hopi people discuss this relationship. She added that, although some of the figure complexes are fairly well understood, spirals are not because the spiral operates as a general symbol, which can have a large number of related meanings. Studies of associated figures, Schaafsma agreed, would be valuable in this respect.

In response to a question from the audience concerning the origin of the spiral, Schaafsma replied that these figures are not found with any frequency in early rock art. The spiral complex was prominent between AD 900 or 1000 and 1300, and there is some possibility that the spiral motif originated in Mexico.

Seowtewa, who has a great interest in the rock art surrounding the Pueblo of Zuni, remarked that the early rock art can be explained with reference to nature. As an example, he cited snake petroglyphs, which may be related to early spring or summer when the snakes come out of their places of hibernation. He added that, at Zuni, some rock art images are said to be clan symbols; others represent the migration of the Zuni people before they found the Center Place. He echoed the sentiments of the panelists that this subject of the meaning of rock art symbols is confusing for all involved.

An additional perspective was supplied by Zeilik's comment that J. W. Fewkes, while working at Hopi, pointed out a connection between the Mesoamerican water design and the spiral symbol that was being used on late nineteenth century Hopi pottery. Zeilik added that, although there is evidence that certain symbols were regarded by the historical Pueblos as sun symbols, there is no indication as to which symbols, if any, might have constituted sun symbols in prehistoric rock art elements.

Considering contemporary interpretations, Eggan reported that, according to a Hopi chief from Second Mesa, shrines on the outer boundary of the Hopi reservation are marked by spirals on the rock walls. He concluded that for the modern-day Hopi, at least, spirals are markers of ancestral boundaries.

Citing suggestions made by Sofaer and her colleagues that the rock slabs at Fajada Butte had been moved to create the "light dagger" effect, Carlson asked the Prestons, who have extensively documented rock art sites that demonstrate light/shadow effects, "Do these effects involve natural configurations of rock, or is there evidence that the rocks were in any way shaped, modified, or moved?"

R. Preston responded that the sites he and his wife have documented seem to be composed of natural formations in which the only human intervention was placement of the petroglyphs.

Zeilik remarked that most of the sites intentionally created to display light/shadow effects date to prehistoric times, so perhaps the connection made in the past between spirals and the sun is no longer made--an indication of changing symbolism in the Southwest.

R. Preston replied that perhaps there are fewer known cases of relatively recent rock art spirals interacting with light images because spirals no longer appear with much frequency in contemporary Southwestern rock art. P. Schaafsma commented that the spirals serving as Hopi boundary markers mentioned by Eggan may be prehistoric.

Reiterating the problem of rock art symbolism, A. Preston said that one cannot make simple assumptions about the meaning of symbols because sites can be multifunctional and because symbols at such sites can be embedded in a complex of interrelated meanings.

Citing research with contemporary Zuni interpretations of rock art surrounding the pueblo, Young added that there is seldom complete agreement about meanings of images, particularly for the older rock art figures; therefore, one cannot generally claim to have a cultural consensus about the meanings of symbols in rock art, nor is that meaning necessarily oriented toward astronomical themes, even for images that outsiders consider to be sun symbols. Young suggested that just as Jojola described the many variants of myths and stories that abound in Pueblo culture, so too are there variants in interpretations or stories about rock art. She criticized researchers who only listen to the versions that fit their archaeoastronomical theories, thereby attempting to freeze and solidify meanings that are fluid and complex.

Sofaer remarked that perhaps unwarranted assumptions are made about what constitutes a sun symbol. For instance, a rock art image that interacts with a sun dagger is not necessarily a sun symbol in Pueblo thought.

At Hopi, Sekaquaptewa said, these spiral figures are associated with migratory legends and initiations, which involve salt-gathering trips to the Grand Canyon and subsequent inscription of clan symbols on certain rocks. He suggested that spirals may indicate the passage of time as well as the marking of spiritual and physical centers. Spirals that are not perfect spirals, he said, may be a philosophical expression of the hardships of life.

LIMITATIONS OF ETHNOGRAPHIC ANALOGY

Since much of the archaeoastronomical research on prehistoric sites in the Southwest has relied on analogies to historical and contemporary Pueblo societies, Judge raised the crucial issue of the validity of such analogies, cautioning that attention must be paid to the tremendous time span between the Chacoan societies and the contemporary Pueblos. He argued that one cannot simply transfer the past onto the present, or vice versa, especially if one considers that the Chacoan system may have been a unique manifestation that is not comparable to modern or historical Pueblo cultures. For these reasons, Judge criticized over-reliance on contemporary Puebloan interpretations of rock art as indices of the meaning of prehistoric rock art.

Jojola carried this argument one step further, stating that the modern Pueblo society is perhaps more complex than is usually realized by those who, in attempting to map the present onto the past, treat Puebloan society as a unified whole. Jojola emphasized that Pueblo societies are not homogeneous, but are characterized by different perspectives revealed in the different stories by which phenomena are explained. For example, Jojola cited the symbol of the snake or rattlesnake, which at his own Pueblo of Isleta is said to depict an epidemic of rattlesnakes. It is neither regarded as being related to the solstices nor considered to be a means by which to measure the seasons. In sum, Jojola noted that there are, of course, common elements in contemporary Puebloan societies, but there are also fundamental differences that must be taken into account, rendering it extremely difficult to map them onto an equally complex past.

As an alternative to attempts to draw ethnographic parallels between the present and the past, Williamson suggested that the relationships among rock art styles within the context of Chaco Canyon itself should be analyzed.

Judge responded that, although Col. Jim Bain has completed the rock art survey of Chaco Canyon, a systematic analysis such as those conducted by P. Schaafsma in other areas of the Southwest has not yet been completed. In addition, Judge commented that it may not be especially notable that light-shadow-spiral phenomena in Chaco Canyon and elsewhere marked the solstices, since several of the papers presented at this symposium argued that the timing of the solstices was public knowledge. He suggested that, instead, the rock art in Chaco Canyon should be considered in relation to the predictive dimension, the esoteric knowledge of astronomy.

CEREMONIALISM, SYMBOLISM, AND FUNCTIONALISM

Referring to an essential dimension of Native American astronomical practice, Eggan suggested that the panelists should turn to a consideration of the title of the symposium, which includes the words astronomy and ceremony. He commented that the discussion up to this point had focused on scientific astronomy--measurement and techniques--and that, although the panelists had come to the conclusion that the Anasazi were good observers, they had not considered the fact that such data were gathered in order to construct a cosmology. Eggan urged the panelists to be aware that this cosmology had been developed in order to establish a system of prediction and control, and that Puebloan religion, of which cosmology is an integral part, is designed to keep the system going in an uncertain, unpredictable environment. Eggan emphasized that one cannot deal with Puebloan societies only in terms of strict scientific methodology. Echoing earlier statements by both Seowtewa and Sekaquaptewa, Eggan stated that the spiritual side as well as the scientific side of Puebloan cosmology must be taken into account by archaeoastronomers.

These remarks sparked a discussion of the functional and symbolic aspects of Puebloan cosmology. It was mentioned, for example, that although the four posts that support the roof of the kiva are said to represent the four directions, this number of pillars may simply have been the most efficient. Williamson suggested, however, that the use of the four posts in the kiva could be attributed to both functional and mythological considerations. He noted that one should not look only for literal explanations: sometimes there is a disjuncture between what is appropriate in an ideal sense and what is actually done. He cited the historical Pawnee earthlodges as an example in which one might find six or eight posts, only four of which were described as actually supporting the roof. Williamson stressed that it is necessary to be aware of both the functional and the symbolic attributes of the structures under investigation.

ROCK ART AND THE CRAB NEBULA SUPERNOVA (AD 1054)

Carlson then directed the discussion to a consideration of possible representations of the Crab Nebula supernova in Southwestern rock art, asking whether it could be decided, one way or another, if this event was recorded at certain rock art sites. In addition, he asked, if it could be demonstrated that this event was recorded, what would the implications of such a record be?

Zeilik responded that the rock art figure below Peñasco Blanco in Chaco Canyon, which some scholars have suggested represents the Crab Nebula supernova, could simply be a depiction of Venus next to the crescent moon. Although such rock art figures could have marked a sun-watching station at this site, Zeilik's research indicates that, for an observer standing below the rock art figures, there is not enough irregularity along the horizon for the site to have served the purpose of establishing a good horizon calendar. Zeilik pointed out that poor potential of this site for horizon-watching does not mean that the rock art represents the Crab Nebula supernova. Although many modern astronomers have adopted this interpretation, the meaning of this site in the context of Anasazi culture is still unknown.

Jojola added that within his culture such unusual celestial phenomena as comets are often interpreted as forewarnings of disaster. It is unlikely, therefore, that a phenomenon such as the Crab Nebula supernova would have been documented in rock art. A problem archaeoastronomers face, he reiterated, is that they are trying to interpret a symbol in isolation, outside its appropriate cultural context. Instead, they should be aware that a symbol is only part of a larger cosmological context and should be interpreted only in terms of that association.

Aveni suggested that the interpretation of this site by modern astronomers might be another example of what he had described earlier as an overarching belief in cultural transcendence and, consequently, the inability to see beyond the perspective of one's own culture. The problem arises, Aveni said, in assuming that because one culture places a high value on recording spectacularand cataclysmic events, another culture would as well. In contrast, he pointed out that the Anasazi may instead have recorded only those phenomena that most concerned them--those that were recurrent, part of a continuing, observable cycle.

Reyman brought a new perspective to the discussion by mentioning that a Pueblo friend he once took to the site at Peñasco Blanco commented that perhaps the archaeoastronomers were confused by a fundamental misunderstanding about sun-watching stations. He warned against confusing the place from which one makes the observation with the sunwatcher's station. According to Reyman, he meant that the sunwatcher goes to certain places on a ritual trek, as part of the obligations of this office, that have no astronomical implications at all. The markings at Peñasco Blanco might, therefore, serve simply to indicate that this was one of the stops on the sunwatcher's ritual trek.

COMPLEXITY OF CHACOAN SOCIETY

A consideration of the Chacoan road system was the next topic Carlson raised for discussion. Wilcox addressed the issue by asking Judge how much labor per mile would have been necessary to build the roads. Judge responded that a tremendous amount of labor had been invested. Although the roads were clearly engineered, he pointed out that this does not mean that they were not the result of the labor of a peasant class. As an example, Judge suggested that the energy invested by members of San Juan Pueblo in one year in ditch maintenance equalled the amount of energy per year that was required to build all of the Chacoan ruins. He added that there is clear evidence of cultural complexity in Chacoan society, and this complexity may have entailed stratified labor investment--that is, a system in which one class of people performed the labor and another class directed the work.

Judge's comments led a member of the audience to ask whether this meant that there was a group of engineers associated with the priestly class at Chaco who built the roads and the buildings. Reyman responded that a priestly class had not been identified for the Chacoan culture, let alone an engineering group per se. By way of comparison, he added that although there is evidence of communal labor employed by the historical Pueblos to construct certain buildings there is no evidence of a special group of engineers.

Some specialized knowledge must have been required, Carlson countered, because many of the buildings at Chaco Canyon were very carefully aligned and structured in a manner that could not have been accidental.

To clarify, McCluskey stressed that evidence of precise engineering or precise astronomical observation does not necessarily imply a specialized class of engineers or astronomers. He suggested that one could accept expertise in construction, in sun watching, and in predicting astronomical phenomena, within the context of a generalized class of elders, without the need to posit the existence of specialists within this class who were engineers or astronomers.

In response to a question raised by Judge concerning the method for laying out an accurate east-west line, Williamson suggested the following method (illustrated in Figure 13.1): "Place a stick perpendicular to the earth in the center of a

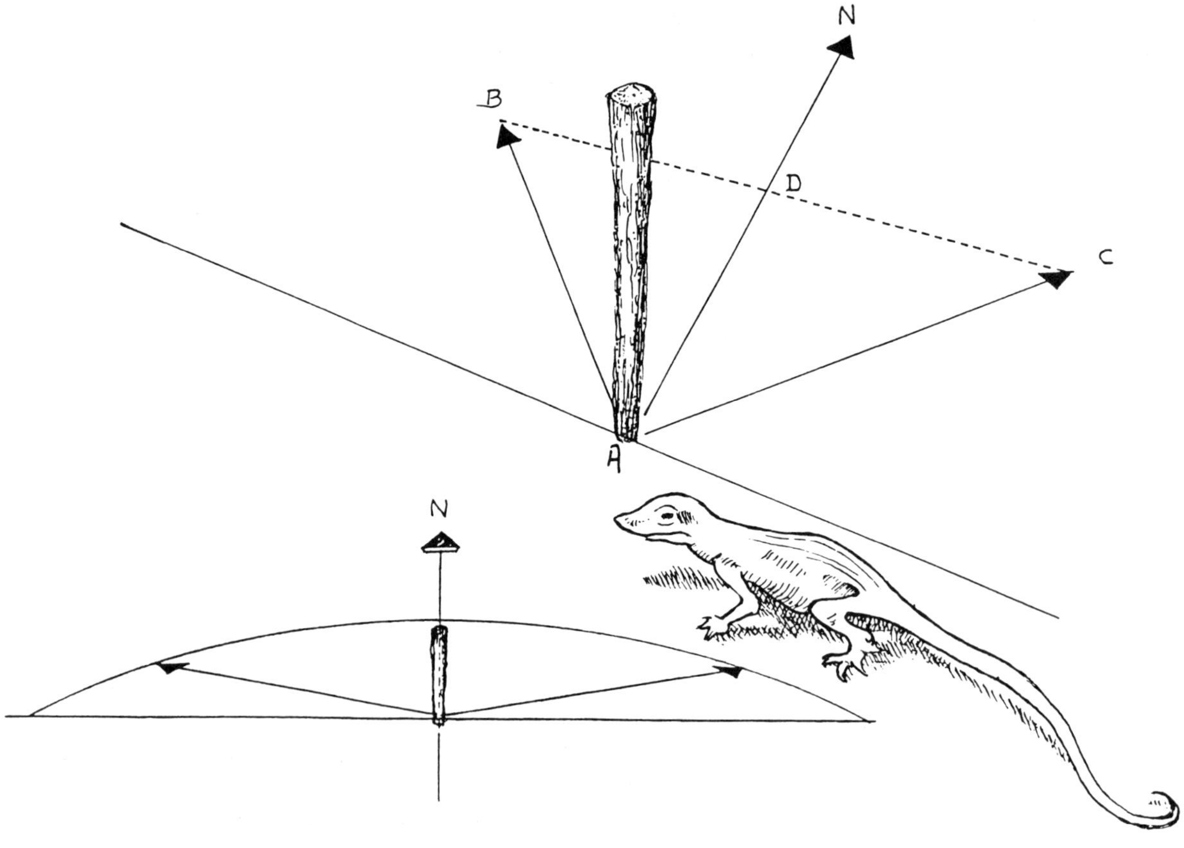

Figure 13.1. Simple tools that could be used for determining cardinal alignments. Illustration by Snowden Hodges (from Living the Sky: The Cosmos of the American Indian, by Ray A. Williamson, p. 144. Houghton Mifflin, Boston, 1984)

smoothed area. Mark the endpoints of the stick's shadow throughout the day and, picking out two shadows on either side of noon that are of the same length, draw a line connecting their endpoints (B and C). The bisector of line BC drawn through point A (line AD) is a north-south line. Line BC is an east-west line." This is simple geometry, Williamson remarked, but the result turns out to be very accurate.

Sofaer asked if this method could be used to pick out the minute around noon. Williamson commented that it could, but that one must also consider the question of intentionality--why would anyone want to do that? Sofaer speculated that the east-west axis at Pueblo Bonito might have been regarded by the Anasazi as the middle of space and that the equinox rise and set phenomena at Fajada Butte--the constant movement of shadows across petroglyphs--might indicate that Fajada symbolized a center in space and time. She added that shadow markings on some of the petroglyphs at Fajada occur within a few minutes of noon. According to Sofaer, all of this might suggest that "the center of the day of the center of the year was of some importance along with the east-west alignment at Bonito."

Carlson remarked that there seems to be a consensus that simple technology--probably not a corps of engineers--was required to perform such tasks as aligning the roads, building the kivas, and marking the light and shadow phenomena at Chaco Canyon. McCluskey, Reyman, Judge, and Wilcox concurred with this, while the other panelists seemed to offer tacit agreement.

Considering the reasons behind the construction of such structures, Wilcox remarked on the difficulty, when looking to the past, of distinguishing between the certain and the possible. Symbolism, he said, seems to fall into the realm of the possible; therefore, those who assign symbolic as well as functional attributes to these prehistoric building plans may be imposing their own perceptions onto the data. For example, Wilcox stated that it is almost certain that the buildings at Casa Grande were designed--that the builders had a conception of the structure as a whole before they started building it. He pointed out, however, that although this could be a step closer to symbolism, it is not a necessary step. On the other hand, he countered, it would be an error in judgement to assume that because something was not needed, it therefore did not exist.

McCluskey reiterated his earlier point that, although the Chacoan people had specialized architectural knowledge, this does not imply that they needed a specialized architectural class to maintain that knowledge.

Jojola interjected that in his village of Isleta, the women are the architects. This point was not pursued, but it deserves further attention.

Wilcox concluded that the question of whether or not buildings at Casa Grande were designed to reflect cultural symbolism probably cannot be resolved, but he felt that similar cosmologies existed among the Hohokam at Casa Grande and in other Southwestern cultures. He said the awareness of cultural/cosmological similarities between different groups was valuable to the archaeoastronomical endeavor--it encourages one to look for new relationships in the archaeological record, a new kind of data that goes beyond empirical fact.

A critical comment made by an audience member was that archaeoastronomers do not seem to follow the methodology of deductive reasoning, or to test their hypotheses. In partial response, Wilcox suggested that although archaeoastronomers are aware of problems such as these, there may as yet be no viable means by which to test their hypotheses.

In sum, the panel served as a format for discussing critical and controversial issues of recent archaeoastronomical research in the Southwest. Clearly, much remains to be done in this area, and further research ought to be guided by more critical scholarship, interdisciplinary dialogue, and input from the Puebloan peoples whose astronomical practices are so often cited. A true synthesis of the various claims regarding astronomy and ceremonialism is not yet possible, although the panel, and the symposium as a whole, provided an important step toward that goal.

ACKNOWLEDGMENTS

My thanks to Anthony F. Aveni, Jonathan Reyman, Curt Schaafsma, David Wilcox, and Ray A. Williamson for sending me their notes on the panel discussion when it became clear that sections of the tapes were inaudible. I am grateful to Deborah Fant for her expert editorial advice.